MODULAR SERIES
ON SOLID STATE DEVICES

VOLUME I
Semiconductor Fundamentals
Second Edition

Foreword

Solid state devices have attained a level of sophistication and economic importance far beyond the highest expectations of their inventors. By continually offering better performing devices at lower cost per unit, the electronics industry has penetrated markets never before addressed. A necessary prerequisite to sustaining this growth initiative is an enhanced understanding of the internal workings of solid state devices by modern electronic circuit and systems designers. This is essential because system, circuit, and IC layout design procedures are being merged into a single function. Considering the present and projected needs, we have established this series of books (the Modular Series) to provide a strong intuitive and analytical foundation for dealing with solid state devices.

Volumes I through IV of the Modular Series are written for junior, senior, or possibly first-year graduate students who have had at least an introductory exposure to electric field theory. Emphasis is placed on developing a fundamental understanding of the internal workings of the most basic solid state device structures. With some deletions, the material in the first four volumes is used in a one-semester, three credit-hour, junior-senior course in electrical engineering at Purdue University. The material in each volume is specifically designed to be presented in 10 to 12 fifty-minute lectures.

The volumes of the series are relatively independent of each other, with certain necessary formulas repeated and referenced between volumes. This flexibility enables one to use the volumes sequentially or in selected parts, either as the text for a complete course or as supplemental material. It is also hoped that students, practicing engineers, and scientists will find the series useful for individual instruction, whether it be for reference, review, or home study.

A number of the standard texts on devices have been written like encyclopedias, packed with information, but with little thought as to how the student learns or reasons. Texts that are encyclopedic in nature are often difficult for students to read and may even present barriers to understanding. By breaking the material into smaller units of information, and by writing *for students*, we have hopefully constructed volumes which are truly readable and comprehensible. We have also sought to strike a healthy balance between the presentation of basic concepts and practical information.

Problems which are included at the end of each chapter constitute an important component of the learning program. Some of the problems extend the theory presented in the text or are designed to reinforce topics of prime importance. Others are numerical problems which provide the reader with an intuitive feel for the typical size of key parameters. When approximations are stated or assumed, the student will then have confidence that cited quantities are indeed orders of magnitude smaller than others. The end-of-chapter problems range in difficulty from very simple to quite challenging. In the second edition we have added a new feature — worked problems or *exercises*. The exercises are collected in Appendix A and are referenced from the appropriate point within the text. The exercises are similar in nature to the end-of-chapter problems. Finally, Appendix B contains volume-review problem sets and answers. These sets contain short answer, test-like, questions which could serve as a review or as a means of self-evaluation.

Reiterating, the emphasis in the first four volumes is on developing a keen understanding of the internal workings of the most basic solid state device structures. However, it is our hope that the volumes will also help (and perhaps motivate) the reader to extend his knowledge — to learn about the many more devices already in use and to even seek information about those presently in research laboratories.

Prof. Gerold W. Neudeck
Prof. Robert F. Pierret
Purdue University
School of Electrical Engineering
W. Lafayette, IN 47907

Contents

Introduction

The primary purpose of this volume is to concisely present and examine those terms, concepts, equations, models, etc., that are routinely employed in describing the operational behavior of solid state devices. In essence, the reader will be exposed to the basics of the language spoken by the solid state device specialist, a language that must be mastered if the reader and the device specialist are to communicate efficiently. The first chapter in the volume provides a general physical description of semiconductors, surveying, for example, their elemental makeup, atomic arrangements, and similar topics. The second chapter presents a detailed description of the current carrying entities inside a semiconductor under "rest" (equilibrium) conditions. The third and final chapter examines what happens to the current carrying entities when the semiconductor is electrically perturbed, and how the resulting situation is described mathematically in terms of observables and known parameters.

The reader should be forewarned that the subject matter covered in this volume is moderately challenging and at times somewhat conceptual in nature. However, in this work every effort has been made to be as lucid as possible without jeopardizing the integrity of the topic. One final comment: It should be noted that the sum total of the subject matter has been reduced to the absolute minimum required for a basic understanding of devices.

1 / Semiconductors — General Information

1.1 GENERAL MATERIAL PROPERTIES

The vast majority of all solid state devices on the market today are fabricated from a class of materials known as semiconductors. It is therefore appropriate that we begin the discussion by examining the general nature of semiconducting materials.

1.1.1 Composition

Table 1.1 lists the atomic compositions of semiconductors which are likely to be encountered in the device literature. As noted, the semiconductor family of materials includes the elemental semiconductors Si and Ge, compound semiconductors such as GaAs and CdTe, and alloys like $Al_xGa_{1-x}As$.[†] Due in large part to the advanced state of its fabrication technology, Si is far and away the most important of the semiconductors, totally dominating the present commercial market. The vast majority of integrated

Table 1.1 Semiconductor Materials.

General Classification	Specific Examples
(1) Elemental Semiconductors	Si, Ge
(2a) III-V Compounds	AlP, AlAs, AlSb, GaN, GaP, GaAs, GaSb, InP, InAs, InSb
(2b) II-VI Compounds	ZnO, ZnS, ZnSe, ZnTe, CdS, CdSe, CdTe, HgS
(3) Alloys	$Al_xGa_{1-x}As$, $GaAs_{1-x}P_x$, $Hg_{1-x}Cd_xTe$, $Ga_xIn_{1-x}As_{1-y}P_y$

[†] The x (or y) in alloy formulas is a fraction lying between 0 and 1. $Al_{0.3}Ga_{0.7}As$ would indicate a material with 3 Al and 7 Ga atoms per every 10 As atoms.

circuits and most devices are made with this material. GaAs, exhibiting superior electrical properties and special optical properties, is finding increased usage in a number of applications, including high-speed IC's, and will someday share a significant portion of the commercial marketplace. The remaining semiconductors are primarily utilized in specialized applications which are most often of an optoelectronic nature. Given its present position of dominance, we will tend to focus our attention on Si in the subsequent discussion; where feasible, GaAs will be given comparable consideration.

Although the number of semiconducting materials is reasonably large, the list is actually quite limited considering the total number of elements and possible combinations of elements. Note that, referring to the abbreviated periodic chart of the elements in Table 1.2, only a certain group of elements and elemental combinations typically give rise to semiconducting materials. Specifically, all of the semiconductors listed in Table 1.1 are composed of elements appearing in Column IV of the Periodic Table, or are a combination of elements in Periodic Table columns an equal distance to either side of Column IV. The Column III element Ga plus the Column V element As yields the III-V compound semiconductor GaAs; the Column II element Cd plus the Column VI element Te yields the II-VI compound semiconductor CdTe; the fractional combination of the Column III elements Al and Ga plus the Column V element As yields the alloy semiconductor $Al_xGa_{1-x}As$. This very general property is related to the chemical bonding in semiconductors, where, on the average, there are four valence electrons per atom.

1.1.2 Purity

As will be explained in Chapter 2, extremely minute traces of impurity atoms called "dopants" can have a drastic effect on the electrical properties of semiconductors. For this reason, the compositional purity of semiconductors must be very carefully controlled and, in fact, modern semiconductors are some of the purest solid materials in existence. In Si, for example, the unintentional content of dopant atoms is routinely less than one atom per 10^9 Si atoms. To assist the reader in attempting to comprehend this incredible level of purity, let us suppose a forest of maple trees was planted from coast to coast, border to border, at 50-ft centers across the United States (including Alaska). An impurity level of one part per 10^9 would correspond to finding about 25 crabapple trees in the maple tree forest covering the United States! It should be emphasized that the cited material purity refers to *unintentional* undesired impurities. Typically, dopant atoms at levels ranging from one part per 10^8 to one impurity atom per 10^3 semiconductor atoms will be *purposely* added to the semiconductor to control its electrical properties.

1.1.3 Structure

The spatial arrangement of atoms within a material plays an important role in determining the precise properties of the material. As shown schematically in Fig. 1.1, the

Table 1.2   Abbreviated Periodic Chart of the Elements.

II	III	IV	V	VI
4 Be	5 B	6 C	7 N	8 O
12 Mg	13 Al	14 Si	15 P	16 S
30 Zn	31 Ga	32 Ge	33 As	34 Se
48 Cd	49 In	50 Sn	51 Sb	52 Te
80 Hg	81 Tl	82 Pb	83 Bi	84 Po

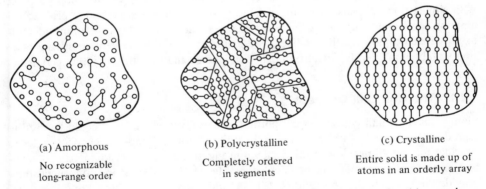

(a) Amorphous

No recognizable
long-range order

(b) Polycrystalline

Completely ordered
in segments

(c) Crystalline

Entire solid is made up of
atoms in an orderly array

Fig. 1.1   General classification of solids based on the degree of atomic order: (a) amorphous, (b) polycrystalline, and (c) crystalline.

atomic arrangement within a solid causes it to be placed into one of three broad classifications; namely, amorphous, polycrystalline, or crystalline. An amorphous solid is a material in which there is no recognizable long-range order in the positioning of atoms within the material. The atomic arrangement in any given section of an amorphous material will look different from the atomic arrangement in any other section of the material. Crystalline solids lie at the opposite end of the "order" spectrum; in a crystalline material the atoms are arranged in an orderly three-dimensional array. Given any section of a crystalline material, one can readily reproduce the atomic arrangement in any other section of the material. Polycrystalline solids comprise an intermediate case in which the solid is composed of crystalline subsections that are disjointed or misoriented relative to each other.

Upon examining the many solid state devices in existence, one readily finds examples of all three structural forms. An amorphous-Si thin-film transistor is used as the switching element in liquid crystal displays; polycrystalline Si gates are employed in Metal-Oxide-Semiconductor Field-Effect Transistors (MOSFET's). In the vast majority of devices, however, the active region of the device is situated within a *crystalline* semiconductor. The overwhelming number of devices fabricated today employ *crystalline* semiconductors.

1.2 CRYSTAL STRUCTURE

The discussion at the end of the preceding section leads nicely into the topic of this section. Since in-use semiconductors are typically crystalline in form, it seems reasonable to seek out additional information about the crystalline state. Our major goal here is to present a more detailed picture of the atomic arrangement within the principal semiconductors. To achieve this goal, we first examine how one goes about describing the spatial positioning of atoms within crystals. Next, a bit of "visualization" practice with simple three-dimensional lattices (atomic arrangements) is presented prior to examining semiconductor lattices themselves. The section concludes with an introduction to the related topic of Miller indices. Miller indices are a convenient shorthand notation widely employed for identifying specific planes and directions within crystals.

1.2.1 The Unit Cell Concept

Simply stated, a unit cell is a small portion of any given crystal that could be used to reproduce the crystal. To help establish the unit cell (or building-block) concept, let us consider the two-dimensional lattice shown in Fig. 1.2(a). In order to describe this lattice or to totally specify the physical characteristics of the lattice, one need only provide the unit cell shown in Fig. 1.2(b). As indicated in Fig. 1.2(c), the original lattice can be readily reproduced by merely duplicating the unit cell and stacking the duplicates next to each other in an orderly fashion.

In dealing with unit cells there often arises a misunderstanding, and hence confusion, relative to two points. First of all, unit cells are not necessarily unique. The unit cell shown in Fig. 1.2(d) is as acceptable as the Fig. 1.2(b) unit cell for specifying the original lattice of Fig. 1.2(a). Second, a unit cell need not be primitive (the smallest unit cell possible). In fact, it is usually advantageous to employ a slightly larger unit cell with orthogonal sides instead of a primitive cell with nonorthogonal sides. This is especially true in three dimensions where noncubic unit cells are quite difficult to describe and visualize.

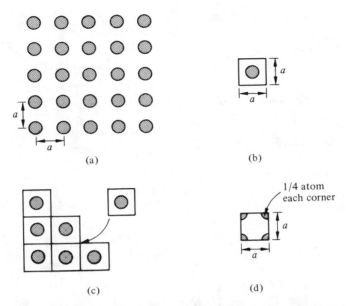

(a)   (b)

(c)   (d)

Fig. 1.2 Introduction to the unit cell method of describing atomic arrangements within crystals. (a) Sample two-dimensional lattice. (b) Unit cell corresponding to the part (a) lattice. (c) Reproduction of the original lattice. (d) An alternative unit cell.

1.2.2 Simple 3-D Unit Cells

Semiconductor crystals are three-dimensional and are therefore described in terms of three-dimensional (3-D) unit cells. In Fig. 1.3(a) we have pictured the simplest of all three-dimensional unit cells; namely, the simple cubic unit cell. The simple cubic cell is an equal-sided box or cube with an atom positioned at each corner of the cube. The simple cubic lattice associated with this cell is constructed in a manner paralleling the two-dimensional case. In doing so, however, it should be noted that only 1/8 of each corner atom is actually *inside* the cell, as pictured in Fig. 1.3(b). Duplicating the Fig. 1.3(b) cell and stacking the duplicates like blocks in a nursery yields the simple cubic lattice.

Figures 1.3(c) and 1.3(d) display two common 3-D unit cells that are somewhat more complex but still closely related to the simple cubic cell. The unit cell of Fig. 1.3(c) has an atom added at the center of the cube; this configuration is appropriately called the body centered cubic (bcc) unit cell. The face centered cubic (fcc) unit cell of Fig. 1.3(d) contains an atom at each face of the cube in addition to the atoms at each corner. (Note, however, that only one-half of each face atom actually lies inside the fcc unit cell.) Whereas the simple cubic cell contains one atom (1/8 of an atom at

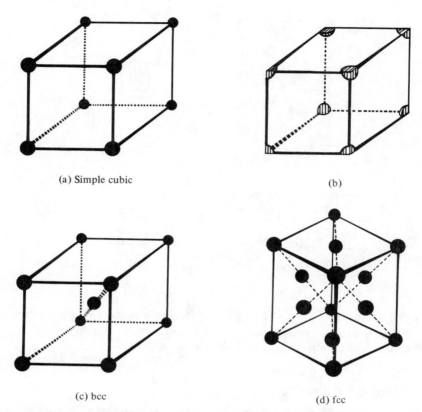

(a) Simple cubic   (b)

(c) bcc   (d) fcc

Fig. 1.3 Simple three-dimensional unit cells. (a) Simple cubic unit cell. (b) Pedantically correct simple cubic unit cell including only the fractional portion (1/8) of each corner atom actually within the cell cube. (c) Body centered cubic unit cell. (d) Face centered cubic unit cell. [(a), (c), and (d) from *Physics of Solids* by C. A. Wert and R. M. Thomson, © 1964 by McGraw-Hill Book Co., New York.]

each of the eight cube corners), the somewhat more complex bcc and fcc cells contain two and four atoms, respectively. The reader should verify these facts and visualize the lattices associated with the bcc and fcc cells.

1.2.3 Semiconductor Lattices

We are finally in a position to supply details relative to the positioning of atoms within the principal semiconductors. In Si (and Ge) the lattice structure is described by the unit cell pictured in Fig. 1.4(a). The Fig. 1.4(a) arrangement is known as the *diamond lattice* unit cell because it also characterizes diamond, a form of the Column IV ele-

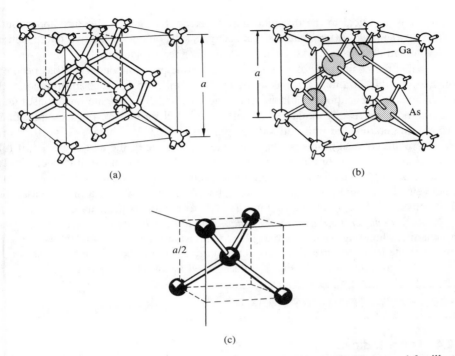

Fig. 1.4   (a) Diamond lattice unit cell. (b) Zincblende lattice unit cell (GaAs used for illustration). (c) Enlarged top corner of the part (a) diamond lattice emphasizing the four-nearest-neighbor bonding within the structure. The cube side length, a, is 5.43 Å and 5.65 Å at room temperature for Si and GaAs, respectively.

ment carbon. Examining the diamond lattice unit cell, we note that the cell is cubic, with atoms at each corner and at each face of the cube similar to the fcc cell. The interior of the Fig. 1.4(a) cell, however, contains four additional atoms. One of the interior atoms is located along a cube body diagonal exactly one-quarter of the way down the diagonal from the top front left-hand corner of the cube. The other three interior atoms are displaced one-quarter of the body diagonal length along the previously noted body diagonal direction from the front, top, and left-side face atoms, respectively. Although it may be difficult to visualize from Fig. 1.4(a), the diamond lattice can also be described as nothing more than two interpenetrating fcc lattices. The corner and face atoms of the unit cell can be viewed as belonging to one fcc lattice, while the atoms totally contained within the cell belong to the second fcc lattice. The second lattice is displaced one-quarter of a body diagonal along a body diagonal direction relative to the first fcc lattice.

Most of the III-V semiconductors, including GaAs, crystallize in the *zincblende* structure. The zincblende lattice, typified by the GaAs unit cell shown in Fig. 1.4(b), is essentially identical to the diamond lattice, except that lattice sites are apportioned equally between two different atoms. Ga occupies sites on one of the two interpenetrating fcc sublattices; arsenic (As) populates the other fcc sublattice.

Now that the positioning of atoms within the principal semiconductors has been established, the question may arise as to the practical utilization of such information. Although several applications could be cited, geometrical-type calculations constitute a very common and readily explained use of the unit cell formalism. For example, in Si at room temperature the unit cell side length (a) is 5.43 Å (where 1 Å = 10^{-8} cm). Since there are eight Si atoms per unit cell and the volume of the unit cell is a^3, it follows that there are $8/a^3$ or almost exactly 5×10^{22} atoms/cm^3 in the Si lattice. Similar calculations could be performed to determine atomic radii, the distance between atomic planes, etc. For the purposes of the development herein, however, a major reason for the discussion of semiconductor lattices was to establish that, as emphasized in Fig. 1.4(c), *atoms in the diamond and zincblende lattices have four nearest neighbors.* The chemical bonding (or crystalline glue) within the major semiconductors is therefore dominated by the attraction between any given atom and its four closest neighbors. This is an important fact that should be filed away for future reference.

> **SEE EXERCISE 1.1–APPENDIX A**

1.2.4 Miller Indices

Single crystals of silicon used in device processing normally assume the form and shape pictured in Fig. 1.5. This single crystal of silicon, better known as a Si wafer, is commonly ~0.5 mm thick and 100 to 150 mm (4 to 6 in.) in diameter, with at least

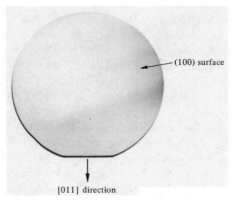

Fig. 1.5 Single crystal silicon wafer with a (100) surface and a [011] flat. Si wafers with (100) and (111) surfaces are standard; other orientations can be obtained only by special order. The figure is primarily intended to dramatize the utility of Miller indices.

one surface carefully polished and etched to yield a damage-free, mirrorlike finish. Of particular interest here is the fact that the wafer's surface is also carefully preoriented to lie along a specific crystallographic plane, and a "flat" is ground along the periphery of the wafer to identify a reference direction within the surface plane. Precise surface orientation is critical in certain device processing steps and directly affects the characteristics exhibited by a number of devices; the "flat" is routinely employed to orient device arrays on the wafer so as to achieve high yields during device separation. The point we wish to make is that specification of crystallographic planes and directions is of practical importance. Miller indices, exemplified by the (100) plane and [011] direction designations used in Fig. 1.5, constitute the conventional means of identifying planes and directions within a crystal.

The Miller indices for any given plane of atoms within a crystal are obtained by following a straightforward four-step procedure. The procedure is detailed below, along with the simultaneous sample indexing of the plane shown in Fig. 1.6(a).

Indexing Procedure	Sample Implementation
After setting up coordinate axes along the edges of the unit cell, note where the plane to be indexed intercepts the axes. Divide each intercept value by the unit cell length along the respective coordinate axis. Record the resulting normalized (pure-number) intercept set in the order x, y, z.	$1, 2, 3$
Invert the intercept values; that is, form $[1/\text{intercept}]$s.	$1, \frac{1}{2}, \frac{1}{3}$
Using an appropriate multiplier, convert the $1/\text{intercept}$ set to the smallest possible set of whole numbers.	$6, 3, 2$
Enclose the whole-number set in curvilinear brackets.	(632)

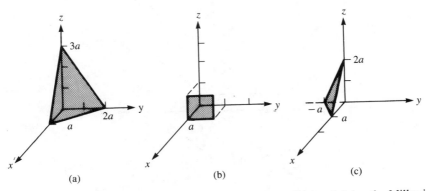

Fig. 1.6 Sample cubic crystal planes. (a) The (632) plane used in explaining the Miller indexing procedure. (b) The (100) plane. (c) The (2$\bar{2}$1) plane.

To complete the description of the plane-indexing procedure, the user should also be aware of the following special facts:

i) If the plane to be indexed is parallel to a coordinate axis, the "intercept" along the axis is taken to be at infinity. Thus, for example, the plane shown in Fig. 1.6(b) intercepts the coordinate axes at 1, ∞, ∞, and is therefore a (100) plane.

ii) If the plane to be indexed has an intercept along the negative portion of a coordinate axis, a minus sign is placed *over* the corresponding index number. Thus the Fig. 1.6(c) plane is designated a ($2\bar{2}1$) plane.

iii) Referring to the diamond lattice of Fig. 1.4(a), note that the six planes passing through the cube faces contain identical atom arrangements; that is, because of crystal symmetry, it is impossible to distinguish between the "equivalent" (100), (010), (001), ($\bar{1}$00), ($0\bar{1}0$), and ($00\bar{1}$) planes; or, it is impossible to distinguish between {100} planes. A group of equivalent planes is concisely referenced in the Miller notation through the use of { } braces.

The Miller indices for *directions* are established in a manner analogous to the well-known procedure for finding the components of a vector. First set up a vector of arbitrary length in the direction of interest. Next, decompose the vector into its components by noting the projections of the vector along the coordinate axes. Using an appropriate multiplier, convert the component values into the smallest possible whole-number set. This of course changes the length of the original vector but not its direction. Finally, enclose the whole-number set in brackets. Square brackets, [ ], are used to designate specific directions within a crystal; triangular brackets, ⟨ ⟩, designate an equivalent set of directions. Sample direction vectors and their corresponding Miller indices are displayed in Fig. 1.7. A summary of the Miller notation is given in Table 1.3.

Fig. 1.7 Sample direction vectors and their corresponding Miller indices.

Table 1.3 Miller Notation Summary.

Convention	Interpretation
(hkl)	Crystal plane
$\{hkl\}$	Equivalent planes
$[hkl]$	Crystal direction
$\langle hkl \rangle$	Equivalent directions

In the foregoing discussion we presented the procedure for progressing from a given plane or direction in a crystal to the corresponding Miller indices. More often than not, one is faced with the inverse process — visualizing the crystalline plane or direction corresponding to a given set of indices. Fortunately, one seldom encounters other than low-index planes and directions such as (111), (110), [001], etc. Thus it is possible to become fairly adept at the inverse process by simply memorizing the orientations for planes and directions associated with small-number indices. It is also helpful to note that, *for cubic crystals, a plane and the direction normal to the plane have precisely the same indices* — e.g., the [110] direction is normal to the (110) plane. Of course, any plane or direction can always be deduced by reversing the indexing procedure.

SEE EXERCISE 1.2–APPENDIX A

1.3 CRYSTAL GROWTH

1.3.1 Obtaining Ultrapure Si

Given the obvious general availability and widespread use of semiconductor materials, particularly Si, it is reasonable to wonder about the origin of the single-crystal Si used in modern-day device production. Could it be that Si is readily available in sandstone deposits? — No. Perhaps, as a sort of by-product, Si single crystals come from South African diamond mines? — Wrong again. As suggested recently in a low-budget science fiction movie, maybe Si is scooped from the ocean bottom by special submarines? — Sorry, no. Although Si is the second most abundant element in the earth's crust and a component in numerous compounds, chief of which are silica (impure SiO_2) and the silicates (Si + O + another element), silicon never occurs alone in nature as an element. Single-crystal Si used in device production, it turns out, is a man-made material.

Given the above introduction, it should be clear that the initial step in producing device-quality silicon must involve separating Si from its compounds and purifying the separated material. The ingenious separation and purification process that has evolved

Fig. 1.8 Summary of the process employed to produce ultrapure silicon.

is summarized in Fig. 1.8. Low-grade silicon or ferrosilicon is first produced by heating silica with carbon in an electric furnace. The carbon essentially pulls the oxygen away from the impure SiO_2 (i.e., *reduces* the SiO_2), leaving behind impure Si. Next, the ferrosilicon is chlorinated to yield $SiCl_4$ or $SiHCl_3$, both of which are liquids at room temperature. Although appearing odd at first glance, the liquefaction process is actually a clever maneuver. Whereas solids are very difficult to purify, a number of standard procedures are available for purifying liquids. An ultrapure $SiCl_4$ (or $SiHCl_3$) is the result after multiple distillation and other liquid purification procedures. Lastly, the high-purity halide is chemically reduced, yielding the desired ultrapure elemental silicon. This is accomplished, for example, by heating $SiCl_4$ in a hydrogen atmosphere $[SiCl_4 + 2H_2 \rightarrow 4HCl + Si]$.

1.3.2 Single Crystal Formation

Although ultrapure, the silicon derived from the separation and purification process as just described is not a single crystal, but is instead polycrystalline. Additional processing is therefore required to form the large single crystals used in device fabrication. The most commonly employed method yielding large single crystals of Si is known as the Czochralski pulled method. In this method the ultrapure polycrystalline silicon is placed in a quartz crucible and heated in an inert atmosphere to form a melt, as shown in Fig. 1.9. A small single crystal, or Si "seed" crystal, with the normal to its bottom face carefully aligned along a predetermined direction (typically a $\langle 111 \rangle$ or $\langle 100 \rangle$ direction), is then clamped to a metal rod and dipped into the melt. Once thermal equilibrium is established, the temperature of the melt in the vicinity of the seed crystal is reduced, and silicon from the melt begins to freeze out onto the seed crystal, the added material being a structurally perfect extension of the seed crystal. Subsequently, the seed crystal is slowly rotated and withdrawn from the melt; this permits more and more silicon to freeze out on the bottom of the growing crystal. The resulting large, cylindrically shaped single crystal of silicon, also known as an ingot (see Fig. 1.10), is typically 4 to 6 in. in diameter and 1 to 2 m in length. The Si wafers used in device process-

Fig. 1.9 Typical silicon crystal-pulling apparatus. (a) Photograph of a modern computer-controlled Czochralski crystal puller. (b) Schematic drawing of the puller. [(a) Courtesy of Cybeq Systems, a business unit of Siltec Corp. (b) From W. Zuhlehner and D. Huber, "Czochralski Grown Silicon," *Crystals 8,* Springer-Verlag, Berlin (1982).]

ing (Fig. 1.5) are ultimately produced by cutting the ingot into thin sections using a diamond-edged saw.

1.4 SUMMARY

This chapter provides basic information about semiconductors in general and silicon in particular. Examining the elemental compositions of semiconductors, one finds that there are, on average, four valence electrons per atom. Moreover, device-quality semiconductors are typically single crystals of high compositional purity. Si crystallizes in the diamond lattice, GaAs crystallizes in the zincblende lattice, and the lattice atoms in both cases have four nearest neighbors. Miller indices, introduced herein as an adjunct

Fig. 1.10   Silicon ingots and crystal growth materials. Pictured from left to right on the table are: Back row—three quartz crucibles in varying sizes, graphite crucible holder, and crucible filled with polycrystalline Si chunks, bag of polycrystalline Si, portion of a 4″ Si ingot. Second through fourth rows—3″, 1½″, and 2″ Si ingots, respectively. Fifth row—4″ Si ingot, portion of a 4″ Si crystal, saw blade used in slicing the ingots into wafers. Front row—package of dopant material and a seed crystal. (Photograph courtesy of Delco Electronics Div., GMC, Kokomo, Indiana.)

to the crystal structure discussion, are the accepted means for identifying planes and directions within a crystal. Finally, large single crystals of device-grade Si are commonly produced by the Czochralski pulled method.

PROBLEMS

1.1 The GaAs unit cell is pictured in Fig. 1.4(b). Describe (or sketch) the unit cell for $Al_{0.5}Ga_{0.5}As$.

1.2 Given a lattice constant of $a = 5.65 \times 10^{-8}$ cm, determine the number of Ge atoms per cm^3 at room temperature.

1.3 In terms of the lattice constant a, what is the distance between nearest-neighbor atoms in

(a) a bcc lattice?

(b) an fcc lattice?

1.4 The surface of a Si wafer is a (100) plane.

(a) Sketch the placement of Si atoms on the surface of the wafer.

(b) Determine the number of atoms per cm^2 at the surface of the wafer.

(c) and (d) Repeat (a) and (b) taking the surface of the Si wafer to be a (110) plane.

1.5 Record all intermediate steps in answering the following questions.

(a) As shown in Fig. P1.5(a), a crystalline plane has intercepts of $1a$, $3a$, and $1a$ on the x, y, and z axes, respectively. a is the cubic cell side length.

  (i) What is the Miller index notation for the plane?

  (ii) What is the Miller index notation for the direction normal to the plane?

(b) Assuming the crystal structure to be cubic, determine the Miller indices for (i) the plane and (ii) the vector pictured in Fig. P1.5(b).

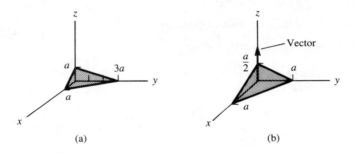

(a)   (b)

Fig. P1.5

1.6 Assuming a cubic crystal system, make a sketch of the following planes.

       (a) (001),   (b) (111),   (c) (123),   (d) ($\bar{1}$10),

       (e) (010),   (f) ($\bar{1}\bar{1}\bar{1}$),   (g) (221),   (h) (0$\bar{1}$0).

1.7 Assuming a cubic crystal system, use an appropriately directed arrow to identify each of the following directions.

       (a) [010],   (b) [101],   (c) [00$\bar{1}$],   (d) [111],

       (e) [001],   (f) [110],   (g) [0$\bar{1}$0],   (h) [123].

1.8 Identify two crystalline directions in a cubic crystal which are perpendicular to

(a) the [100] direction,

(b) the [111] direction.

NOTE: The cosine of the angle θ between two arbitrary directions, $[h_1k_1l_1]$ and $[h_2k_2l_2]$, in a cubic crystal is

$$\cos(\theta) = \frac{h_1h_2 + k_1k_2 + l_1l_2}{[(h_1^2 + k_1^2 + l_1^2)(h_2^2 + k_2^2 + l_2^2)]^{1/2}}$$

1.9 Given a cubic lattice, indicate how many equivalent planes [parts (a) to (c)] or equivalent directions [parts (d) to (f)] are associated with each of the following designations:

$$\text{(a) } \{100\}, \quad \text{(b) } \{110\}, \quad \text{(c) } \{111\},$$

$$\text{(d) } \langle 100 \rangle, \quad \text{(e) } \langle 110 \rangle, \quad \text{(f) } \langle 111 \rangle.$$

1.10 Treating atoms as rigid spheres with radii equal to one-half the distance between nearest neighbors, show that the ratio of the volume occupied by the atoms to the total available volume in the various crystal structures is:

(a) $\pi/6$ or 52% for the simple cubic lattice,

(b) $\sqrt{3}\,\pi/8$ or 68% for the body centered cubic lattice,

(c) $\sqrt{2}\,\pi/6$ or 74% for the face centered cubic lattice,

(d) $\sqrt{3}\,\pi/16$ or 34% for the diamond lattice.

2 / Carrier Modeling

Carriers are the entities that transport charge from place to place inside a material and hence give rise to electrical currents. In everyday life the most commonly encountered type of carrier is the electron, the subatomic particle responsible for charge transport in metallic wires. Within semiconductors one again encounters the familiar electron, but there is also a second equally important type of carrier—the hole. Electrons and holes are the focal point of this chapter, wherein we examine carrier related concepts, models, properties, and terminology.

Although reminders will be periodically interjected, it should be emphasized from the very start that the development throughout this chapter assumes *equilibrium* conditions exist within the semiconductor. "Equilibrium" is the term used to describe the unperturbed state of a system. Under equilibrium conditions there are no external voltages, magnetic fields, stresses, or other perturbing forces acting on the semiconductor. All observables are invariant with time. This "rest" condition, as we will see, provides an excellent *frame of reference*. Being able to characterize the semiconductor under equilibrium conditions permits one to extrapolate and ascertain the semiconductor's condition when a perturbation has been applied.

As a final introductory comment, the reader should be forewarned that a number of formulas and facts will be presented without justification in this chapter. We would like to provide a complete discussion, properly developing every concept and deriving every formula. Unfortunately, this is not possible. It is our underlying philosophy, moreover, that being able to derive a result is secondary to knowing how to *interpret* and *use* a result. One can, of course, fill in any information gap through supplemental readings. The interested reader is specifically referred to Volume VI in the Modular Series for additional information.

2.1 THE QUANTIZATION CONCEPT

Instead of attempting to deal immediately with electrons in crystalline Si, where there are 14 electrons per atom and 5×10^{22} atoms/cm^3, we will take a more realistic approach and first establish certain ground rules by examining much simpler atomic systems. We begin with the simplest of all atomic systems, the isolated hydrogen atom. This atom, as the reader may recall from a course in modern physics, was under intense scrutiny at the start of the 20th century. Scientists of that time knew the hydrogen atom contained a negatively charged particle in orbit about a more massive positively charged nucleus. What they could not explain was the nature of the light emitted from the system when the hydrogen atom was heated to an elevated temperature. Specifically, the emitted light was observed at only certain discrete wavelengths; according to the prevailing theory of the time, scientists expected a continuum of wavelengths.

   In 1913 Niels Bohr proposed a solution to the dilemma. Bohr hypothesized that the H-atom electron was restricted to certain well-defined orbits; or, equivalently, Bohr assumed that the orbiting electron could take on only certain values of angular momentum. This "quantization" of the electron's angular momentum was, in turn, coupled directly to energy quantization. As can be readily established, if the electron's angular momentum is assumed to be $\mathbf{n}\hbar$, then

$$E_{\mathrm{H}} = -\frac{m_0 q^4}{2(4\pi\varepsilon_0\hbar\mathbf{n})^2} = -\frac{13.6}{\mathbf{n}^2}\mathrm{eV}, \qquad \mathbf{n} = 1, 2, 3, \ldots \qquad (2.1)$$

where (also see Fig. 2.1) E_{H} is the electron binding energy within the hydrogen atom, m_0 is the mass of a free electron, q is the magnitude of the electronic charge, ε_0 is the permittivity of free space, h is Planck's constant, $\hbar = h/2\pi$, and $\mathbf{n}$ is the energy quantum number or orbit identifier. The *electron volt* (eV) is a unit of energy equal to 1.6×10^{-19} joules. Now, with the electron limited to certain energies inside the hydrogen atom, it follows from the Bohr model that the transition from a higher $\mathbf{n}$ to a lower

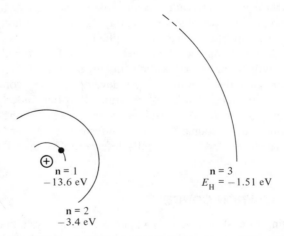

Fig. 2.1   The hydrogen atom—idealized representation showing the first three allowed electron orbits and the associated energy quantization.

n orbit will release quantized energies of light; this explains the observation of emitted light at only certain discrete wavelengths.

For our purposes, the most important idea to be obtained from the Bohr model is that the energy of electrons in atomic systems is restricted to a limited set of values.[†] Relative to the hydrogen atom, the energy level scheme in a multi-electron atom like silicon is, as one might intuitively expect, decidedly more complex. It is still a relatively easy task, however, to describe the salient energy-related features of an isolated silicon atom. As pictured in Fig. 2.2, ten of the 14 Si-atom electrons occupy very deep-lying energy levels and are tightly bound to the nucleus of the atom. The binding is so strong, in fact, that these ten electrons remain essentially unperturbed during chemical reactions or normal atom–atom interactions, with the ten-electron-plus-nucleus combination often being referred to as the "core" of the atom. The remaining four Si-atom electrons, on the other hand, are rather weakly bound and are collectively called the *valence electrons* because of their strong participation in chemical reactions. As emphasized in Fig. 2.2, the four valence electrons, if unperturbed, occupy four of the eight allowed slots (or states) having the next highest energy above the deep-lying core levels. Finally, we should note, for completeness, that the electronic configuration in the 32-electron Ge-atom (germanium being the other elemental semiconductor) is essentially identical to the Si-atom configuration except the Ge-core contains 28 electrons.

2.2 SEMICONDUCTOR MODELS

Building on the information presented in previous sections, we introduce and describe in this section two very important models or visualization aids which are used extensively in the analysis of semiconductor devices. The inclusion of semiconductor mod-

n = 2
8 Electrons

Six allowed levels
at same energy

+14

Two allowed levels
at same energy

s p

n = 1
2 Electrons

n = 3

Fig. 2.2 Schematic representation of an isolated Si atom.

[†]Actually, not only energy but many other observables relating to atomic-sized particles are quantized. An entire field of study, quantum mechanics, has been developed to describe the properties and actions of atomic-sized particles and systems.

els in a chapter devoted to carrier modeling may appear odd at first glance but is actually quite appropriate. We are, in effect, modeling the carrier "container," the semiconductor crystal.

2.2.1 Bonding Model

The isolated Si atom, or a Si atom not interacting with other atoms, was found to contain four valence electrons. Si atoms incorporated in the diamond lattice, on the other hand, exhibit a bonding that involves an attraction between each atom and its four nearest neighbors [refer to Fig. 1.4(c)]. The implication here is that in going from isolated atoms to the collective crystalline state Si atoms come to share one of their valence electrons with each of the four nearest neighbors. This covalent bonding, or equal sharing of valence electrons with nearest neighbors, and the mere fact that atoms in the diamond lattice have four nearest neighbors, give rise to the idealized semiconductor representation, the bonding model, shown in Fig. 2.3. Each circle in the Fig. 2.3 bonding model represents the core of a semiconductor atom, while each line represents a shared valence electron. (There are eight lines connected to each atom because any given atom not only *contributes* four shared electrons but must also *accept* four shared electrons from adjacent atoms.) The two-dimensional nature of the model is, of course, an idealization that facilitates mental visualizations and also makes it much easier to reproduce the model on paper and chalkboards.

Although considerable use will be made of the bonding model in subsequent discussions, it is nevertheless worthwhile at this point to examine sample applications of the model to provide some indication of the model's utility. Two sample applications are presented in Fig. 2.4. In Fig. 2.4(a) we use the bonding model to picture a point defect, a missing atom, in the lattice structure. In Fig. 2.4(b) we visualize the breaking of an atom-to-atom bond and the associated release or freeing of an electron. [Bond breaking (at $T > 0$ K) and defects occur naturally in all semiconductors, and hence (if we may be somewhat overcritical) the basic model of Fig. 2.3 is strictly valid for an entire semiconductor only at $T \simeq 0$ K when the semiconductor in question contains no defects and no impurity atoms.]

Line represents a shared
valence electron

Circle represents the core
of a semiconductor (e.g. Si)
atom

Fig. 2.3   The bonding model.

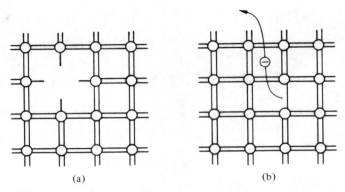

Fig. 2.4 Sample utilization of the bonding model. (a) Visualization of a missing atom or point defect. (b) Breaking of an atom-to-atom bond and freeing of an electron.

2.2.2 Energy Band Model

If our interests were restricted to describing the spatial aspects of events taking place within a semiconductor, the bonding model alone would probably be adequate. Quite often, however, one is more interested in the energy-related aspects of an event. In such instances the bonding model, which says nothing about electron energies, is of little value and the energy band model becomes the primary visualization aid.

Let us begin the conceptual path leading to the energy band model by recalling the situation inside an isolated Si atom. Reviewing the Section 2.1 discussion, ten of the 14 electrons inside an isolated Si atom are tightly bound to the nucleus and are unlikely to be significantly perturbed by normal atom–atom interactions. The remaining four electrons are rather weakly bound and, if unperturbed, occupy four of the eight allowed energy states immediately above the last core level. Moreover, it is implicitly understood that the electronic energy states within a group of Si atoms, say N Si atoms, are all identical — as long as the atoms are isolated, that is, far enough apart so that they are noninteracting.

Given the foregoing knowledge of the isolated atom situation, the question next arises as to whether we can use the knowledge to deduce information about the crystalline state. Assuredly we can omit any further mention of the core electrons because these electrons are not significantly perturbed by normal interatomic forces. The opposite, however, is true of the valence electrons. If N atoms are brought into close proximity (the case in crystalline Si), it is quite reasonable to expect a modification in the energy states of the valence electrons.

The modification in the valence-electron energy states actually known to take place is summarized in Fig. 2.5. In going from N isolated Si atoms to an N-atom Si crystal, exactly half of the allowed states become depressed in energy and half increase in energy. The perturbation, moreover, causes a spread in allowed energies, forming two

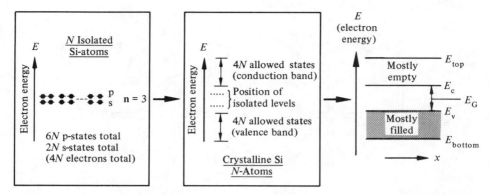

Fig. 2.5 Conceptual development of the energy band model starting with N isolated Si atoms on the left and concluding with a "dressed-up" version of the energy band model on the right.

ranges, or *bands,* of allowed energy states separated by an intervening energy gap. The upper band of allowed states is called the *conduction band;* the lower band of allowed states, the *valence band;* and the intervening energy gap, the *forbidden gap* or *band gap.* In filling the allowed energy band states, electrons, of course, tend to gravitate to the lowest possible energies. Noting that electrons are restricted to single occupancy in allowed states (the Pauli exclusion principle) and remembering that the $4N$ valence band states can just accommodate what were formerly $4N$ valence electrons, we typically find that the valence band is almost completely filled with electrons and the conduction band is all but devoid of electrons. Indeed, the valence band is completely filled and the conduction band completely empty at temperatures approaching $T = 0$ K.

To complete our plausibilization of the energy band model, we need to introduce and utilize one additional fact; namely, unlike the valence electrons in the isolated-atom case, the band electrons in crystalline silicon are not tied to or associated with any one particular atom. True, on the average one will typically find four electrons being shared between any given Si atom and its four nearest neighbors (as in the bonding model). However, the identity of the shared electrons changes as a function of time, with the electrons moving around from point to point in the crystal. In other words, the allowed electronic states are no longer atomic states but are associated with the crystal as a whole; independent of the point examined in a perfect crystal, one sees the same allowed-state configuration. We therefore conclude that for a perfect crystal under equilibrium conditions a plot of the allowed electron energies versus distance along any preselected crystalline direction (always called the x-direction) is as shown on the right-hand side of Fig. 2.5. The cited plot, a plot of allowed electron energy states as a function of position, is the basic energy band model. E_c introduced in the Fig. 2.5 plot is the lowest possible conduction band energy, E_v is the highest possible valence band energy, and $E_G = E_c - E_v$ is the band gap energy.

Fig. 2.6 The energy band diagram — widely employed simplified version of the energy band model.

Finally, Fig. 2.6 displays the form of the energy band model (for a perfect crystal under equilibrium conditions) actually employed in practice. In this widely employed "shorthand" version, the line to indicate the top energy in the conduction band, the line to indicate the bottom energy in the valence band, the cross-hatching to indicate filled states, the labeling of the y- or electron-energy axis, and the labeling of the x- or position axis are all understood to exist implicitly, but are not known explicitly.

2.2.3 Carriers

With the semiconductor models having been firmly established, we are at long last in a position to introduce and to visualize the current-carrying entities within semiconductors. Referring to Fig. 2.7, we note first of all from part (a) that, if there are no broken bonds in the bonding model, or equivalently, if in the energy band model the valence band is completely filled with electrons and the conduction band is devoid of electrons, then there are no carriers. Valence band electrons in the energy band model correspond to shared electrons in the bonding model and these electrons are not involved in charge transport.

The electrons involved in charge transport are visualized in Fig. 2.7(b). When a Si–Si bond is broken and the associated electron is free to wander about the lattice, the released electron is a carrier. Equivalently, in terms of the energy band model, excitation of valence band electrons into the conduction band creates carriers; that is, *electrons in the conduction band are carriers*. Note that the energy required to break a bond in the bonding model and the band gap energy, E_G, are one and the same thing. Likewise, freed bonding-model electrons and conduction band electrons are just different names for the same electrons; in subsequent discussions the word "electrons," when used without a modifier, will be understood to refer to these conduction band electrons.

In addition to releasing an electron, the breaking of a Si–Si bond also creates a missing bond or void in the bonding structure. Thinking in terms of the bonding model, one can visualize the movement of this missing bond from place to place in the lattice as a result of nearby bound electrons jumping into the void [see Fig. 2.7(c)]. Alternatively, one can think in terms of the energy band model where the removal of an electron from the valence band creates an empty state in an otherwise vast sea of filled states. The empty state, like a bubble in a liquid, moves about rather freely in the lat-

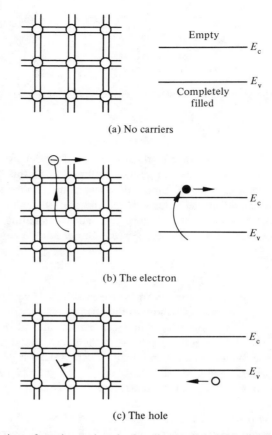

Fig. 2.7 Visualization of carriers using the bonding model (left) and the energy band model (right). (a) No-carrier situation; (b) visualization of an electron; (c) visualization of a hole.

tice because of the cooperative motion of the valence band electrons. What we have been describing, the missing bond in the bonding scheme, the empty state in the valence band, is the second type of carrier found in semiconductors — the hole.

Although the preceding description would appear to belie the fact, the valence band hole is actually an entity on an equal footing with the conduction band electron. As one's familiarity with carrier modeling grows, the comparable status of electrons and holes becomes more and more apparent. Eventually, the reader will come to think of the hole as just another subatomic particle. How can a hole be both a particle and a void? The conceptual dilemma here is a direct result of the almost unavoidable inadequacy of our models. The bonding model in particular is a gross oversimplification of a complex situation and exhibits obvious flaws when examined closely. Truthfully, it is a wonder our models work as well as they do; we are attempting to characterize in everyday terms a system that is beyond the realm of everyday life. Yes, in "reality" the hole is just another subatomic "particle."

2.2.4 Material Classification

Changing the subject somewhat, we would like to conclude this section by citing a very important tie between the band gap of a material, the number of carriers available for transport in a material, and the overall nature of a material. As it turns out, although specifically established for semiconductors, the energy band model of Fig. 2.6 can be applied with only slight modification to all materials. The major difference between materials lies not in the nature of the energy bands, but rather in the magnitude of the energy band gap.

Insulators, as illustrated in Fig. 2.8(a), are characterized by very wide band gaps, with E_G for the insulating materials diamond and SiO_2 being ≈ 5 eV and ≈ 8 eV, respectively. In these wide band gap materials the thermal energy available at room temperature excites very few electrons from the valence band into the conduction band; thus very few carriers exist inside the material and the material is therefore a poor conductor of current. The band gap in metals, by way of comparison, is either very small,

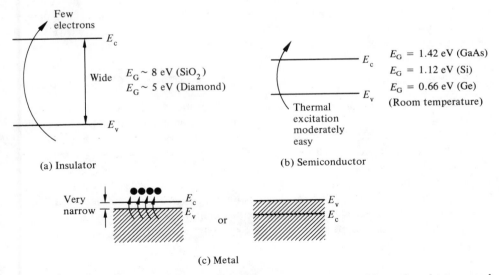

Fig. 2.8 Explanation of the distinction between (a) insulators, (b) semiconductors, and (c) metals using the energy band model.

or no band gap exists at all due to an overlap of the valence and conduction bands [see Fig. 2.8(c)]. There is always an abundance of carriers in metals, and hence metals are excellent conductors. Semiconductors present an intermediate case between insulators and metals. At room temperature, $E_G = 1.42$ eV in GaAs, $E_G = 1.12$ eV in Si, and $E_G = 0.66$ eV in Ge. Thermal energy, by exciting electrons from the valence band into the conduction band, creates a moderate number of carriers in these materials, giving rise in turn to a current-carrying capability intermediate between poor and excellent.

From what we have said, the band gap of a material is clearly an important parameter. Actually, it is perhaps *the* most important material parameter; and the value of the silicon band gap at room temperature ($E_G = 1.12$ eV) should definitely be committed to memory.

2.3 CARRIER PROPERTIES

Having formally introduced the electron and the hole, we next seek to learn as much as possible about the nature of these carriers. In this particular section we examine a collage of carrier related information, information of a general nature including general facts, properties, and terminology.

2.3.1 Charge

Both electrons and holes are charged entities. Electrons are negatively charged, holes are positively charged, and the *magnitude* of the carrier charge, q, is the same for the two types of carriers. In MKS units, $q = 1.6 \times 10^{-19}$ coul. (Please note that, under the convention adopted herein, the electron and hole charges are $-q$ and $+q$, respectively; i.e., the sign of the charge is displayed explicitly.)

2.3.2 Effective Mass

Mass, like charge, is another very basic property of electrons and holes. Unlike charge, however, the carrier mass is not a simple property and cannot be disposed of by simply quoting a number. Indeed, the apparent or effective mass of electrons within a crystal is a function of the semiconductor material (Si, Ge, etc.) and is different from the mass of electrons within a vacuum.

Seeking to obtain insight into the effective mass concept, let us first consider the motion of electrons in a vacuum. If, as illustrated in Fig. 2.9(a), an electron of rest mass m_0 is moving in a vacuum between two parallel plates under the influence of an electric field $\mathcal{E}$, then, according to Newton's second law, the force $\mathbf{F}$ on the electron will be

$$\mathbf{F} = -q\mathcal{E} = m_0 \frac{d\boldsymbol{v}}{dt} \tag{2.2}$$

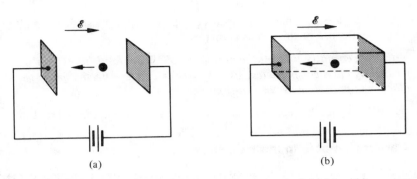

Fig. 2.9 An electron moving in response to an applied electric field (a) within a vacuum, and (b) within a semiconductor crystal.

where v is the electron velocity and t is time. Next consider electrons (conduction band electrons) moving between the two parallel end faces of a semiconductor crystal under the influence of an applied electric field, as envisioned in Fig. 2.9(b). Does Eq. (2.2) also describe the overall motion of electrons within the semiconductor crystal? The answer is clearly *no*. Electrons moving inside a semiconductor crystal will collide with semiconductor atoms, thereby causing a periodic deceleration of the carriers. However, should not Eq. (2.2) apply to the portion of the electronic motion occurring *between* collisions? The answer is again *no*. In addition to the applied electric field, electrons in a crystal are also subject to complex crystalline fields not specifically included in Eq. (2.2).

The foregoing discussion delineated certain important differences between electrons in a crystal and electrons in a vacuum, but left unresolved the equally important question as to how one properly describes the motion of carriers in a crystal. Strictly speaking, the motion of carriers in a crystal can only be described using quantum mechanics, the formalism appropriate for atomic-sized systems. Fortunately, however, if one examines the carrier motion occurring between collisions, the mathematically complex quantum-mechanical formulation simplifies to yield an equation of motion identical to Eq. (2.2), except m_0 is replaced by an effective carrier mass. In other words, for the Fig. 2.9(b) electrons one can write

$$\mathbf{F} = -q\mathcal{E} = m_n^* \frac{d\boldsymbol{v}}{dt} \tag{2.3}$$

where m_n^* is the electron effective mass. A similar equation can be written for holes with $-q \rightarrow q$ and $m_n^* \rightarrow m_p^*$. In each case the internal crystalline fields and quantum-mechanical effects are all suitably lumped into the effective mass factor multiplying $d\boldsymbol{v}/dt$. The significance of this result is that

> both conceptually and mathematically, electrons and holes can be treated as classical particles, as long as the carrier effective mass replaces the particle mass in mathematical relationships.

The carrier effective masses appropriate for Si, Ge, and GaAs are listed in Table 2.1.[†]

2.3.3 Carrier Numbers in Intrinsic Material

The term "intrinsic semiconductor" in common usage refers to an extremely pure semiconductor sample containing an insignificant amount of impurity atoms. More precisely, an intrinsic semiconductor is a semiconductor whose properties are native to the material (that is, not caused by external additives). The number of carriers in an intrinsic semiconductor fits into the scheme of things in that it is an identifiable intrinsic property of the material.

Defining, quite generally,

$$n = \text{number of electrons/cm}^3,$$

$$p = \text{number of holes/cm}^3,$$

existing inside a semiconductor; then, given an intrinsic semiconductor under equilibrium conditions, one finds

$$n = p = n_i \qquad (2.4)$$

Table 2.1 Density of States Effective Masses at 300 K.

Material	m_n^*/m_0	m_p^*/m_0
Si	1.18	0.81
Ge	0.55	0.36
GaAs	0.066	0.52

[†]For the sake of clarity it is sometimes necessary to oversimplify or gloss over complicating issues. The effective mass discussion is a case in point. In reality, the m_n^* appearing in Eq. (2.3) is not a simple scalar quantity, as we have led the reader to believe; nor can one revamp all mathematical relationships using the carrier effective masses listed in Table 2.1. For any given material and carrier type, there are in fact several effective masses encountered in practice. The masses listed in Table 2.1, however, are the most commonly encountered effective masses. Moreover, the precise numerical values of the effective masses are really of secondary importance. Of prime importance is the drastic simplification resulting from the fact that the quantum-mechanical entities known as electrons and holes may be treated, both conceptually and mathematically, as classical particles.

and

$$
\left.
\begin{aligned}
n_i &\simeq 2 \times 10^6/\mathrm{cm}^3 &&\text{in GaAs} \\
&\simeq 1 \times 10^{10}/\mathrm{cm}^3 &&\text{in Si} \\
&\simeq 2 \times 10^{13}/\mathrm{cm}^3 &&\text{in Ge}
\end{aligned}
\right\} \text{at room temperature}
$$

The electron and hole concentrations in an intrinsic semiconductor are equal because carriers within a very pure material can be created only in pairs. Referring to Fig. 2.7, if a semiconductor bond is broken, a free electron and a missing bond or hole are created simultaneously. Likewise, the excitation of an electron from the valence band into the conduction band automatically creates a valence band hole along with the conduction band electron. Also note that the intrinsic carrier concentration, although large in an absolute sense, is relatively small compared with the number of bonds that could be broken. For example, in Si there are 5×10^{22} atoms/cm^3 and four bonds per atom, making a grand total of 2×10^{23} bonds or valence band electrons per cm^3. Since $n_i \simeq 10^{10}/\mathrm{cm}^3$, one finds less than one bond in 10^{13} broken in Si at room temperature. To accurately represent the situation inside intrinsic Si at room temperature, we could cover all the university chalkboards in the world with the bonding model and possibly show only *one* broken bond.

2.3.4   Manipulation of Carrier Numbers — Doping

Doping, in semiconductor terminology, is the addition of controlled amounts of specific impurity atoms with the express purpose of increasing either the electron or the hole concentration. The addition of dopants in controlled amounts to semiconductor materials occurs routinely in the fabrication of almost all semiconductor devices. Common Si dopants are listed in Table 2.2. To increase the electron concentration, one can add either phosphorus, arsenic, or antimony atoms to the Si crystal, with phosphorus being the most commonly employed donor (electron-increasing) dopant. To increase the hole concentration, one adds either boron, gallium, indium, or aluminum atoms to the Si crystal, with boron being the most commonly employed acceptor (hole-increasing) dopant.

Table 2.2   Common Silicon Dopants. Arrows indicate the most widely employed dopants.

Donors (Electron-increasing Dopants)		Acceptors (Hole-increasing Dopants)	
P ←		B ←	
As	Column V elements	Ga	Column III elements
		In	
Sb		Al	

In order to understand how the addition of impurity atoms can lead to a manipulation of carrier numbers, it is important to note that the Table 2.2 donors are all from Column V in the Periodic Table of Elements, while all of the cited acceptors are from Column III in the Periodic Table of Elements. As visualized in Fig. 2.10(a) using the bonding model, when a Column V element with five valence electrons is substituted for a Si atom in the semiconductor lattice, four of the five valence electrons fit snugly into the bonding structure. The fifth donor electron, however, does not fit into the bonding structure and is weakly bound to the donor site. At room temperature this electron is readily freed to wander about the lattice and hence becomes a carrier. Please note that this donation (hence the name "donor") of carrier electrons does not increase the hole concentration. The donor ion left behind when the fifth electron is released cannot move, and there are no broken atom–atom bonds associated with the release of the fifth electron.

The explanation of acceptor action follows a similar line of reasoning. The Column III acceptors have three valence electrons and cannot complete one of the semiconductor bonds when substituted for Si atoms in the semiconductor lattice [see Fig. 2.10(b)]. The Column III atom, however, readily accepts (hence the name "acceptor") an electron from a nearby Si–Si bond, thereby completing its own bonding scheme and in the process creating a hole that can wander about the lattice. Here again there is an increase in only one type of carrier. A negatively charged acceptor site (acceptor atom plus accepted electron) cannot move, and no electrons are released in the hole creation process.

The foregoing bonding-model-based explanation of dopant action is reasonably understandable. There are, nevertheless, a few loose ends. For one, we noted that the fifth donor electron was rather weakly bound and readily freed at room temperature. How does one interpret the relative term "weakly bound"? It takes $\simeq 1$ eV to break Si–Si

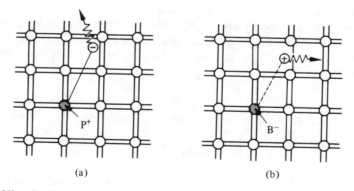

(a)   (b)

Fig. 2.10 Visualization of (a) donor and (b) acceptor action using the bonding model. In (a) the Column V element P is substituted for a Si atom; in (b) the Column III element B is substituted for a Si atom.

bonds and very few of the Si–Si bonds are broken at room temperature. Perhaps "weakly bound" means a binding energy $\simeq 0.1$ eV or less? The question also arises as to how one visualizes dopant action in terms of the energy band model. Both of the cited questions, as it turns out, involve energy considerations and are actually interrelated.

Let us concentrate first on the binding energy of the fifth donor electron. Crudely speaking, the positively charged donor-core-plus-fifth-electron may be likened to a hydrogen atom (see Fig. 2.11). Conceptually, the donor core replaces the hydrogen-atom nucleus and the fifth donor electron replaces the hydrogen-atom electron. In the real hydrogen atom the electron moves of course in a vacuum, can be characterized by the mass of a free electron, and, referring to Eq. (2.1), has a ground-state binding energy of -13.6 eV. In the pseudo-hydrogen atom, on the other hand, the orbiting electron moves through a sea of Si atoms and is characterized by an effective mass. Hence, in the donor or pseudo-atom case, the permittivity of free space must be replaced by the permittivity of Si and m_0 must be replaced by m_n^*. We therefore conclude that the binding energy (E_B) of the fifth donor electron is approximately

$$E_B \simeq -\frac{m_n^* q^4}{2(4\pi K_S \varepsilon_0 \hbar)^2} = \frac{m_n^*}{m_0}\frac{1}{K_S^2}E_{H|n=1} \simeq -0.1 \text{ eV} \qquad (2.5)$$

where K_S is the Si dielectric constant ($K_S \simeq 12$). Actual donor-site binding energies in Si are listed in Table 2.3. The observed binding energies are seen to be in good agreement with the Eq. (2.5) estimate and, confirming the earlier speculation, are roughly $\simeq 1/20$ the Si band gap energy.

Having established the strength of dopant-site bonding, we are now in a position to tackle the problem of how one visualizes dopant action using the energy band model. Working on the problem, we note first of all that when an electron is released from a donor site it becomes a conduction band electron. If the energy absorbed at the donor site is precisely equal to the electron binding energy, the released electron will moreover have the lowest possible energy in the conducting band — namely, E_c. Adding an energy $|E_B|$ to the bound donor-site electron, in other words, *raises* the electron's en-

Fig. 2.11　Pseudo-hydrogen atom model for donor-site bonding.

Table 2.3 Dopant-site Binding Energies.

Donors	$|E_B|$	Acceptors	$|E_B|$
Sb	0.039 eV	B	0.045 eV
P	0.045 eV	Al	0.067 eV
As	0.054 eV	Ga	0.072 eV
		In	0.16 eV

ergy to E_c. Hence, we are led to conclude that the bound electron occupies an allowed electronic level an energy $|E_B|$ below the conduction band edge, or, as visualized in Fig. 2.12, donor sites can be incorporated into the energy band scheme by adding allowed electronic levels at an energy $E_D = E_c - |E_B|$. Note that the donor-site energy level is represented by a set of dashes, instead of a continuous line, because an electron bound to a donor site is localized in space; i.e., a bound electron does not leave the general Δx vicinity of the donor site. The relative closeness of E_D to E_c of course reflects the fact that $E_c - E_D = |E_B| \simeq (1/20)E_G(\text{Si})$.

The actual visualization of dopant action using the energy band model is pictured in Fig. 2.13. Examining the left-hand side of Fig. 2.13(a), one finds all the donor sites filled with bound electrons at temperatures $T \to 0$ K. This is true because very little thermal energy is available to excite electrons from the donor sites into the conduction band at these very low temperatures. The situation changes, of course, as the temperature is increased, with more and more of the weakly bound electrons being donated to the conduction band. At room temperature the ionization of the donor sites is all but total, giving rise to the situation pictured at the extreme right of Fig. 2.13(a). Although we have concentrated on donors, the situation for acceptors is completely analogous. As visualized in Fig. 2.13(b), acceptors introduce allowed electronic levels into the forbidden gap at an energy slightly above the valence band edge. At low temperatures, all of these sites will be empty—there is insufficient energy at temperatures $T \to 0$ K for a valence band electron to make the transition to an acceptor site. Increasing temperature, implying an increased store of thermal energy, facilitates electrons jumping from the valence band onto the acceptor levels. The removal of electrons from the va-

Fig. 2.12 Addition of the $E = E_D$ donor levels to the energy band diagram. Dashes of width Δx emphasize the localized nature of the bound donor-site states.

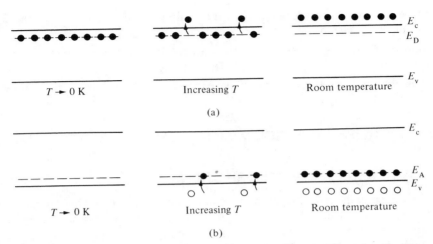

Fig. 2.13   Visualization of (a) donor and (b) acceptor action using the energy band model.

lence band of course creates holes. At room temperature, essentially all of the acceptor sites are filled with an excess electron and an increased hole concentration is effected in the material.

2.3.5   Carrier Related Terminology

Since terminology is often a stumbling block to understanding and since this particular section is replete with specialized terms, it seems appropriate to conclude the section with an overview of carrier related terminology. Approximately one half of the carrier related terms listed below were introduced and defined earlier in this section; the remainder of the terms are being listed here for the first time. All of the terms are widely employed and their definitions should be committed to memory.

   Dopants—specific impurity atoms which are added to semiconductors in controlled amounts for the express purpose of increasing either the electron or the hole concentration.

   Intrinsic semiconductor—undoped semiconductor; extremely pure semiconductor sample containing an insignificant amount of impurity atoms; a semiconductor whose properties are native to the material.

   Extrinsic semiconductor—doped semiconductor; a semiconductor whose properties are controlled by added impurity atoms.

   Donor—impurity atom which increases the electron concentration; n-type dopant.

   Acceptor—impurity atom which increases the hole concentration; p-type dopant.

   n-type material—a donor-doped material; a semiconductor containing more electrons than holes.

p-type material — an acceptor-doped material; a semiconductor containing more holes than electrons.

Majority carrier — the most abundant carrier in a given semiconductor sample; electrons in an *n*-type material, holes in a *p*-type material.

Minority carrier — the least abundant carrier in a given semiconductor sample; holes in an *n*-type material, electrons in a *p*-type material.

SEE EXERCISE 2.1–APPENDIX A

2.4 STATE AND CARRIER DISTRIBUTIONS

Up to this point in the modeling process we have concentrated on carrier properties and information of a conceptual, qualitative or, at most, semiquantitative nature. Practically speaking, there is often a need for more detailed information. For example, most semiconductors are doped, and the precise numerical value of the carrier concentrations inside doped semiconductors is of routine interest. Another property of interest is the distribution of carriers as a function of energy in the respective energy bands. In this section we begin the process of developing a more detailed description of the carrier populations. The development will eventually lead to relationships for the carrier distributions and concentrations within semiconductors under equilibrium conditions.

2.4.1 Density of States

When the energy band model was first introduced in Section 2.2 we indicated that the total number of allowed states in each band was four times the number of atoms in the crystal. Not mentioned at the time was how the allowed states were distributed in energy; that is, how many states were to be found at any given energy in the conduction and valence bands. We are now interested in this energy distribution of states, or *density of states,* as it is more commonly known, because the state distribution is an essential component in determining carrier distributions and concentrations.

To determine the desired density of states it is necessary to perform an analysis based on quantum-mechanical considerations. Herein we will merely summarize the results of the analysis; namely, for energies not too far removed from the band edges, one finds

$$g_c(E) = \frac{m_n^* \sqrt{2m_n^*(E - E_c)}}{\pi^2 \hbar^3}, \qquad E \geq E_c \tag{2.6a}$$

$$g_v(E) = \frac{m_p^* \sqrt{2m_p^*(E_v - E)}}{\pi^2 \hbar^3}, \qquad E \leq E_v \tag{2.6b}$$

where $g_c(E)$ and $g_v(E)$ are the density of states at an energy E in the conduction and valence bands, respectively.

What exactly should the reader know and remember about the cited density of states? Well, for one, it is important to grasp the general density of states concept. The density of states can be likened to the description of the seating in a football stadium, with the number of seats in the stadium a given distance from the playing field corresponding to the number of states a specified energy interval from E_c or E_v. Secondly, the knowledgeable reader would remember the general form of the relationships. As illustrated in Fig. 2.14, $g_c(E)$ is zero at E_c and increases as the square root of energy when one proceeds upward into the conduction band. Similarly, $g_v(E)$ is precisely zero at E_v and increases with the square root of energy as one proceeds downward from E_v into the valence band. Thirdly, one should note that differences between $g_c(E)$ and $g_v(E)$ stem from differences in the carrier effective masses. If m_n^* were equal to m_p^*, the seating (states) on both sides of the football field (the band gap) would be mirror images of each other. Finally, considering closely spaced energies E and $E + dE$ in the respective bands, one can state

$g_c(E)dE$ represents the number of conduction band states/cm^3 lying in the energy range between E and $E + dE$ (if $E \geq E_c$),

$g_v(E)dE$ represents the number of valence band states/cm^3 lying in the energy range between E and $E + dE$ (if $E \leq E_v$).

It therefore follows that $g_c(E)$ and $g_v(E)$ themselves are numbers/unit volume-unit energy, or typically, numbers/cm^3 − eV.

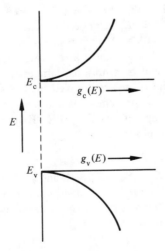

Fig. 2.14 General energy dependence of $g_c(E)$ and $g_v(E)$ near the band edges. $g_c(E)$ and $g_v(E)$ are the density of states in the conduction and valence bands, respectively.

2.4.2 The Fermi Function

Whereas the density of states tells one how many states exist at a given energy E, the Fermi function $f(E)$ specifies how many of the existing states at the energy E will be filled with an electron, or equivalently,

$f(E)$ specifies, under equilibrium conditions, the probability that an available state at an energy E will be occupied by an electron.

Mathematically speaking, the Fermi function is simply a probability distribution function. In mathematical symbols,

$$f(E) = \frac{1}{1 + e^{(E-E_F)/kT}}$$

(2.7)

where

$$E_F = \text{Fermi energy or Fermi level}$$

$$k \ = \text{Boltzmann constant } (k = 8.62 \times 10^{-5} \text{ eV/K})$$

$$T \ = \text{temperature in Kelvin (K)}$$

Seeking insight into the nature of the newly introduced function, let us begin by investigating the Fermi function's energy dependence. Consider first temperatures where $T \rightarrow 0$ K. As $T \rightarrow 0$ K, $(E - E_F)/kT \rightarrow -\infty$ for all energies $E < E_F$ and $(E - E_F)/kT \rightarrow +\infty$ for all energies $E > E_F$. Hence $f(E < E_F) \rightarrow 1/[1 + \exp(-\infty)] = 1$ and $f(E > E_F) \rightarrow 1/[1 + \exp(+\infty)] = 0$. This result is plotted in Fig. 2.15(a) and is simply interpreted as meaning that all states at energies below E_F will be filled and all states at energies above E_F will be empty for temperatures $T \rightarrow 0$ K. In other words, there is a sharp cutoff in the filling of allowed energy states at the Fermi energy E_F when the system temperature approaches absolute zero.

Let us next consider temperatures $T > 0$ K. Examining the Fermi function we make the following pertinent observations.

i) If $E = E_F$, $f(E_F) = 1/2$.

ii) If $E \geq E_F + 3kT$, $\exp[(E - E_F)/kT] \gg 1$ and $f(E) \simeq \exp[-(E - E_F)/kT]$. Consequently, above $E_F + 3kT$ the Fermi function or filled-state probability decays exponentially to zero with increasing energy. Moreover, most states at energies $3kT$ or more above E_F will be empty.

iii) If $E \leq E_F - 3kT$, $\exp[(E - E_F)/kT] \ll 1$ and $f(E) \simeq 1 - \exp[(E - E_F)/kT]$. Below $E_F - 3kT$, therefore, $[1 - f(E)]$, the probability that a given state will be

(a) $T \to 0$ K   (b) $T > 0$ K

Fig. 2.15 Energy dependence of the Fermi function. (a) $T \to 0$ K; (b) generalized $T > 0$ K plot with the energy coordinate expressed in kT units.

empty, decays exponentially to zero with decreasing energy. Most states at energies $3kT$ or more below E_F will be filled.

iv) At room temperature ($T \simeq 300$ K), $kT \simeq 0.026$ eV and $3kT \simeq 0.078$ eV $\ll$ $E_G(\text{Si})$. Compared to the Si band gap, the $3kT$ energy interval that appears prominently in the $T > 0$ K formalism is typically quite small.

The properties cited above are reflected and summarized in the $T > 0$ K Fermi function plot displayed in Fig. 2.15(b).

Before concluding the discussion here, it is perhaps worthwhile to re-emphasize that the Fermi function applies only under equilibrium conditions. The Fermi function, however, is universal in the sense that it applies with equal validity to all materials — insulators, semiconductors, and metals. Although introduced in relationship to semiconductors, the Fermi function is not dependent in any way on the special nature of semiconductors, but is simply a statistical function associated with electrons in general. Finally, the relative positioning of the Fermi energy E_F compared to E_c (or E_v), an item of obvious concern, is treated in subsequent subsections.

SEE EXERCISE 2.2–APPENDIX A

2.4.3 Equilibrium Distribution of Carriers

Having established the distribution of available band states and the probability of filling those states under equilibrium conditions, we can now easily deduce the distribution of carriers in the respective energy bands. To be specific, the desired distribution is obtained by simply multiplying the appropriate density of states by the appropriate occupancy factor — $g_c(E)f(E)$ yields the distribution of electrons in the conduction band and

$g_v(E)[1 - f(E)]$ yields the distribution of holes (unfilled states) in the valence band. Sample carrier distributions for three different assumed positions of the Fermi energy (along with associated energy band diagram, Fermi function, and density of states plots) are pictured in Fig. 2.16.

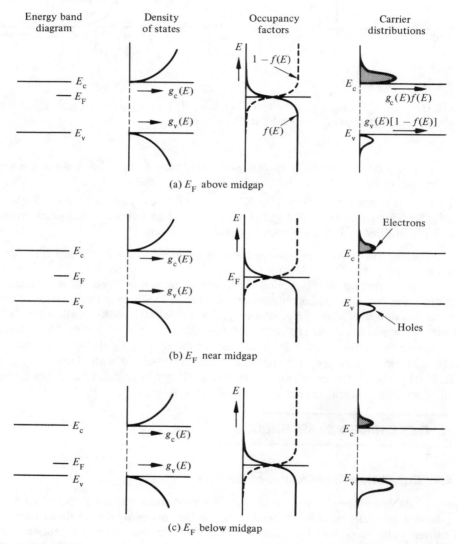

(a) E_F above midgap

(b) E_F near midgap

(c) E_F below midgap

Fig. 2.16 Carrier distributions (not drawn to scale) in the respective bands when the Fermi level is positioned (a) above midgap, (b) near midgap, and (c) below midgap. Also shown in each case are coordinated sketches of the energy band diagram, density of states, and the occupancy factors (the Fermi function and one minus the Fermi function).

Examining Fig. 2.16 we note in general that all carrier distributions are zero at the band edges, reach a peak value very close to E_c or E_v, and then decay very rapidly toward zero as one moves upward into the conduction band or downward into the valence band. In other words, most of the carriers are grouped energetically in the near vicinity of the band edges. Another general point to note is the effect of the Fermi level positioning on the relative magnitude of the carrier distributions. When E_F is positioned in the upper half of the band gap (or higher), the electron distribution greatly outweighs the hole distribution. Although both the filled-state occupancy factor $[f(E)]$ and the empty-state occupancy factor $[1 - f(E)]$ fall off exponentially as one proceeds from the band edges deeper into the conduction and valence bands, respectively, $[1 - f(E)]$ is clearly much smaller than $f(E)$ at corresponding energies if E_F lies in the upper half of the band gap. Lowering E_F effectively slides the occupancy plots downward, giving rise to a nearly equal number of carriers when E_F is at the middle of the gap. Likewise, a predominance of holes results when E_F lies below the middle of the gap. The argument here assumes, of course, that $g_c(E)$ and $g_v(E)$ are of the same order of magnitude at corresponding energies (as is the case for Si and Ge). Also, referring back to the previous subsection, the statements concerning the occupancy factors falling off exponentially in the respective bands are valid provided $E_c - 3kT \geq E_F \geq E_v + 3kT$.

The information just presented concerning the carrier distributions and the relative magnitudes of the carrier numbers finds widespread usage. The information, however, is often conveyed in an abbreviated or shorthand fashion. Figure 2.17, for example, shows a common way of representing the carrier energy distributions. The greatest number of circles or dots are drawn close to E_c and E_v, reflecting the peak in the carrier concentrations near the band edges. The smaller number of dots as one progresses upward into the conduction band crudely models the rapid falloff in the electron density with increasing energy. An extensively utilized means of conveying the relative magnitude of the carrier numbers is displayed in Fig. 2.18. To represent an intrinsic material, a dashed line is drawn in approximately the middle of the band gap and labeled E_i. The near-midgap positioning of E_i, the intrinsic Fermi level, is of course consistent with the previously cited fact that the electron and hole numbers are about equal when E_F is near the center of the band gap. Similarly, a solid line labeled E_F appearing above midgap tells one at a glance that the semiconductor in question is n-type; a solid line labeled E_F appearing below midgap signifies that the semiconductor is p-type. Note finally that the dashed E_i line also typically appears on the energy band diagrams charac-

Fig. 2.17   Schematic representation of carrier energy distributions.

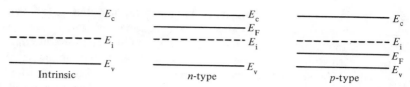

Fig. 2.18 "At a glance" representation of intrinsic (left), n-type (middle), and p-type (right) semiconductor materials using the energy band diagram.

terizing extrinsic semiconductors. The E_i line in such cases represents the expected positioning of the Fermi level *if* the material were intrinsic, and serves as a reference energy level dividing the upper and lower halves of the band gap.

2.5 EQUILIBRIUM CARRIER CONCENTRATIONS

We have arrived at a somewhat critical point in the carrier modeling process. For the most part, this section simply embodies the culmination of our modeling efforts, with working relationships for the equilibrium carrier concentrations being established to reinforce the qualitative carrier information presented in previous sections. Unfortunately, the emphasis on the development of mathematical relationships makes the final assault on the carrier modeling summit the most difficult part of the journey — the reader cannot relax. A comment is also in order concerning the presentation herein of alternative forms for the carrier relationships. The alternative forms can be likened to the different kinds of wrenches used, for example, in home and automobile repairs. The open-end wrench, the box wrench, and the ratchet wrench all serve the same general purpose. In some applications one can use any of the wrenches. In other applications, however, a special situation restricts the type of wrench employed or favors the use of one wrench over another. The same is true of the alternative carrier relationships. Finally, boxes are drawn around expressions that find widespread usage. A single-walled box signifies a moderately important result; a double-walled box, a very important result.

2.5.1 Formulas for *n* and *p*

Since $g_c(E)\,dE$ represents the number of conduction band states/cm³ lying in the E to $E + dE$ energy range, and $f(E)$ specifies the probability that an available state at an energy E will be occupied by an electron, it then follows that $g_c(E)f(E)\,dE$ gives the number of conduction band electrons/cm³ lying in the E to $E + dE$ energy range, and $g_c(E)f(E)\,dE$ integrated over all conduction band energies must yield the total number of electrons in the conduction band. In other words, integration over the equilibrium distribution of electrons in the conduction band yields the equilibrium electron concen-

tration. A similar statement can be made relative to the hole concentration. We therefore conclude

$$n = \int_{E_c}^{E_{top}} g_c(E) f(E) \, dE \tag{2.8a}$$

$$p = \int_{E_{bottom}}^{E_v} g_v(E) [1 - f(E)] \, dE \tag{2.8b}$$

Seeking explicit expressions for the carrier concentrations, let us focus our efforts
on the n-integral. (It will be assumed the reader can perform the p-integral manipulations following an analogous procedure.) Substituting the Eq. (2.6a) expression for
$g_c(E)$ and the Eq. (2.7) expression for $f(E)$ into Eq. (2.8a), one obtains

$$n = \frac{m_n^* \sqrt{2m_n^*}}{\pi^2 \hbar^3} \int_{E_c}^{E_{top}} \frac{\sqrt{E - E_c} \, dE}{1 + e^{(E-E_F)/kT}} \tag{2.9}$$

Now letting

$$\eta = \frac{(E - E_c)}{kT} \tag{2.10a}$$

$$\eta_c = \frac{(E_F - E_c)}{kT} \tag{2.10b}$$

$$E_{top} \to \infty \tag{2.10c}$$

yields

$$n = \frac{m_n^* \sqrt{2m_n^*} (kT)^{3/2}}{\pi^2 \hbar^3} \int_0^\infty \frac{\eta^{1/2} \, d\eta}{1 + e^{\eta - \eta_c}} \tag{2.11}$$

The (2.10c) simplification on the upper integration limit makes use of the fact that the
integrand in question falls off rapidly with increasing energy and is essentially zero for
energies only a few kT above E_c. Hence, extending the upper limit to ∞ has a totally
negligible effect on the value of the integral.

Even with the cited simplification, the Eq. (2.11) integral cannot be expressed in a
closed form containing simple functions. The integral itself is in fact a tabulated function which can be found in a variety of mathematical references. Identifying

$$F_{1/2}(\eta_c) \equiv \int_0^\infty \frac{\eta^{1/2} \, d\eta}{1 + e^{\eta - \eta_c}}, \quad \text{the Fermi-Dirac integral of order } \tfrac{1}{2} \tag{2.12}$$

and also defining

$$N_C = 2\left[\frac{2\pi m_n^* kT}{h^2}\right]^{3/2}, \quad \text{the "effective" density of conduction band states} \quad (2.13a)$$

$$N_V = 2\left[\frac{2\pi m_p^* kT}{h^2}\right]^{3/2}, \quad \text{the "effective" density of valence band states} \quad (2.13b)$$

one obtains

$$n = N_C \frac{2}{\sqrt{\pi}} F_{1/2}(\eta_c) \qquad (2.14a)$$

and, by analogy,

$$p = N_V \frac{2}{\sqrt{\pi}} F_{1/2}(\eta_v) \qquad (2.14b)$$

where $\eta_v \equiv (E_v - E_F)/kT$.

The Eq. (2.14) relationships are a very general result, valid for any conceivable positioning of the Fermi level. The constants N_C and N_V are readily calculated; at 300 K, $N_{C,V} = (2.509 \times 10^{19}/\text{cm}^3)\,(m_{n,p}^*/m_0)^{3/2}$. The value of the Fermi integral can be obtained from available tables, plots, or by direct computation. The general-form relationships, nonetheless, are admittedly cumbersome and inconvenient to use in routine analyses. Fortunately, simplified closed-form expressions do exist which can be employed in the vast majority of practical problems. To be specific, if E_F is restricted to values $E_F \leq E_c - 3kT$, then $1/[1 + \exp(\eta - \eta_c)] \simeq \exp[-(\eta - \eta_c)]$ for all $E \geq E_c(\eta \geq 0)$, and

$$F_{1/2}(\eta_c) = \frac{\sqrt{\pi}}{2} e^{(E_F - E_c)/kT} \qquad (2.15a)$$

Likewise, if $E_F \geq E_v + 3kT$, then

$$F_{1/2}(\eta_v) = \frac{\sqrt{\pi}}{2} e^{(E_v - E_F)/kT} \qquad (2.15b)$$

It therefore follows that, if $E_v + 3kT \leq E_F \leq E_c - 3kT$,

$$n = N_C e^{(E_F - E_c)/kT} \tag{2.16a}$$

$$p = N_V e^{(E_v - E_F)/kT} \tag{2.16b}$$

The mathematical simplification leading to Eqs. (2.16) is equivalent to approximating the filled-state and empty-state occupancy factors by exponential functions — an approximation earlier shown to be valid provided E_F was somewhere in the band gap no closer than $3kT$ to either band edge. Whenever E_F is confined, as noted, to $E_v + 3kT \leq E_F \leq E_c - 3kT$, instead of continually repeating the E_F restriction, the semiconductor is simply said to be *nondegenerate*. Whenever E_F lies in the band gap closer than $3kT$ to either band edge or actually penetrates one of the bands, the semiconductor is said to be *degenerate*. These very important terms are also defined pictorially in Fig. 2.19.

2.5.2   Alternative Expressions for *n* and *p*

Although in closed form, the Eq. (2.16) relationships are not in the simplest form possible, and, more often than not, it is the simpler alternative form of these relationships which one encounters in device analyses. The alternative-form relationships can be obtained by recalling that E_i, the Fermi level for an intrinsic semiconductor, lies close to midgap, and hence Eqs. (2.16) most assuredly apply to an intrinsic semiconductor. If this be the case, then specializing Eqs. (2.16) to an intrinsic semiconductor, i.e., setting $n = p = n_i$ and $E_i = E_F$, one obtains

$$n_i = N_C e^{(E_i - E_c)/kT} \tag{2.17a}$$

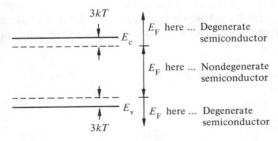

Fig. 2.19 Definition of degenerate/nondegenerate semiconductors.

and

$$n_i = N_V e^{(E_V - E_i)/kT} \qquad (2.17b)$$

Solving Eqs. (2.17) for N_C and N_V yields

$$N_C = n_i e^{(E_c - E_i)/kT} \qquad (2.18a)$$

$$N_V = n_i e^{(E_i - E_V)/kT} \qquad (2.18b)$$

Finally, eliminating N_C and N_V in the original Eq. (2.16) relationships using Eqs. (2.18) gives

$$n = n_i e^{(E_F - E_i)/kT} \qquad (2.19a)$$

$$p = n_i e^{(E_i - E_F)/kT} \qquad (2.19b)$$

Like Eqs. (2.16), the Eq. (2.19) expressions are valid for any semiconductor in equilibrium whose doping is such as to give rise to a nondegenerate positioning of the Fermi level. However, whereas two constants and three energy levels appear in the original relationships, only one constant and two energy levels appear in the alternative relationships. Because of their symmetrical nature, the alternative expressions are also easier to remember, requiring merely an interchange of E_F and E_i in going from the n-formula to the p-formula.

2.5.3 n_i and the np Product

As can be inferred from its appearance in Eqs. (2.19), the intrinsic carrier concentration can figure prominently in the quantitative calculation of the carrier concentrations. Continuing to establish pertinent carrier concentration relationships, we next interject considerations specifically involving this important material parameter.

First of all, if the corresponding sides of Eqs. (2.17a) and (2.17b) are multiplied together, one obtains

$$n_i^2 = N_C N_V e^{-(E_c - E_V)/kT} = N_C N_V e^{-E_G/kT} \qquad (2.20)$$

or

$$n_i = \sqrt{N_C N_V}\, e^{-E_G/2kT} \qquad (2.21)$$

Equation (2.21) expresses n_i as a function of known quantities and may be used to compute n_i at a specified temperature or as a function of temperature. Numerical values for the intrinsic carrier concentration in Si and Ge at room temperature were cited previously; the best available plots of n_i as functions of temperature in Si, Ge, and GaAs are displayed in Fig. 2.20.[†]

A second very important n_i-based relationship follows directly from Eqs. (2.19). Multiplying the corresponding sides of Eqs. (2.19a) and (2.19b) together yields

$$np = n_i^2 \qquad\qquad (2.22)$$

Although appearing trivial, the np product relationship [Eq. (2.22)] often proves to be extremely useful in practical computations. If one of the carrier concentrations is known, the remaining carrier concentration is readily determined using Eq. (2.22)— provided of course that the semiconductor is in equilibrium and nondegenerate.

2.5.4 Charge Neutrality Relationship

The relationships established to this point are devoid of an explicit dependence on the dopant concentrations introduced into a semiconductor. It is the charge neutrality relationship that provides the general tie between the carrier and dopant concentrations. The charge neutrality relationship will be our final addition to the rather impressive arsenal of basic concentration-related expressions.

To establish the charge neutrality relationship, let us consider a *uniformly doped* semiconductor, a semiconductor where the number of dopant atoms/cm^3 is the same everywhere. Systematically examining little sections of the semiconductor and assuming equilibrium conditions, one must invariably find that each and every section of the material is charge-neutral, that is, it exhibits no net charge. If this were not the case, electric fields would exist inside the semiconductor. The electric fields in turn would give rise to carrier motion and associated currents — a situation totally inconsistent with the assumed equilibrium conditions. There are, however, charged entities inside all semiconductors. Electrons, holes, donor sites that have donated an electron to the conduction band (referred to as ionized donor sites), and ionized (negatively charged) acceptor sites may all exist simultaneously inside any given semiconductor. For the uniformly doped material to be everywhere charge-neutral clearly requires

$$\frac{\text{charge}}{\text{cm}^3} = qp - qn + qN_D^+ - qN_A^- = 0 \qquad\qquad (2.23)$$

[†]Substitution of the parameters provided herein into Eq. (2.21) will not yield n_i values precisely as graphed in Fig. 2.20. The primary reason for this discrepancy is the weak but nonnegligible temperature dependence of the effective masses and E_G.

Si	
T (°C)	n_i (cm^{-3})
0	1.04×10^9
5	1.70×10^9
10	2.71×10^9
15	4.28×10^9
20	6.64×10^9
25	1.02×10^{10}
30	1.54×10^{10}
35	2.29×10^{10}
40	3.37×10^{10}
45	4.90×10^{10}
50	7.06×10^{10}
300 K	1.18×10^{10}

GaAs	
T (°C)	n_i (cm^{-3})
0	1.02×10^5
5	1.89×10^5
10	3.45×10^5
15	6.15×10^5
20	1.08×10^6
25	1.85×10^6
30	3.13×10^6
35	5.20×10^6
40	8.51×10^6
45	1.37×10^7
50	2.18×10^7
300 K	2.25×10^6

Fig. 2.20 Intrinsic carrier concentrations in Ge, Si, and GaAs as a function of temperature.

or

$$p - n + N_D^+ - N_A^- = 0 \qquad\qquad (2.24)$$

where, by definition,

N_D^+ = number of ionized (positively charged) donor sites/cm^3,

N_A^- = number of ionized (negatively charged) acceptor sites/cm^3.

As previously discussed, there is sufficient thermal energy available in a semiconductor at room temperature to ionize almost all of the shallow-level donor and acceptor sites. Defining

N_D = total number of donor atoms or sites/cm^3,

N_A = total number of acceptor atoms or sites/cm^3,

and setting

$$N_D^+ = N_D$$
$$N_A^- = N_A$$

one then obtains

$$p - n + N_D - N_A = 0 \qquad \begin{array}{l}\text{assumes total ionization}\\\text{of dopant sites}\end{array} \qquad (2.25)$$

Equation (2.25) is the standard form of the charge neutrality relationship.

2.5.5 Carrier Concentration Calculations

We are finally in a position where we can calculate the carrier concentrations in a uniformly doped semiconductor under equilibrium conditions. In the computations to be presented we specifically make the assumptions of NONDEGENERACY (allowing us to use the np product relationship) and TOTAL IONIZATION of the dopant sites. Note that n_i, which appears in the np product expression, has been calculated and plotted and must be considered a known quantity. Likewise, N_A and N_D, which appear in the charge neutrality relationship, are typically controlled and determined experimentally, and should also be considered known quantities. The only other symbols used in the two equations are n and p. Thus, under the cited assumptions of nondegeneracy and total ionization of dopant sites, we have two equations and two unknowns from which n and p can be deduced.

Starting with the np product expression, one can write

$$p = n_i^2/n \tag{2.26}$$

Eliminating p in Eq. (2.25) using Eq. (2.26) gives

$$n_i^2/n - n + N_D - N_A = 0 \tag{2.27}$$

or

$$n^2 - n(N_D - N_A) - n_i^2 = 0 \tag{2.28}$$

Solving the quadratic equation for n then yields

$$n = \frac{N_D - N_A}{2} + \left[\left(\frac{N_D - N_A}{2} \right)^2 + n_i^2 \right]^{1/2} \tag{2.29a}$$

and

$$p = \frac{n_i^2}{n} = \frac{N_A - N_D}{2} + \left[\left(\frac{N_A - N_D}{2} \right)^2 + n_i^2 \right]^{1/2} \tag{2.29b}$$

Only the plus roots have been retained in Eqs. (2.29) because physically the carrier concentrations must be greater than or equal to zero.

Equations (2.29) are general-case solutions. In the vast majority of practical computations it is possible to simplify these equations prior to substituting in numerical values for N_D, N_A, and n_i. Special cases of specific interest are considered below.

(1) Intrinsic Semiconductor ($N_A = 0$, $N_D = 0$). With $N_A = 0$ and $N_D = 0$, Eqs. (2.29) simplify to $\boxed{n = n_i}$ and $\boxed{p = n_i}$. $n = p = n_i$ is of course the expected result for the equilibrium carrier concentrations in an intrinsic semiconductor.

(2) Doped Semiconductor where either $N_D - N_A \simeq N_D \gg n_i$ *or* $N_A - N_D \simeq N_A \gg n_i$. This is the special case of greatest practical interest. The unintentional doping levels in Si are such that the controlled addition of dopants routinely yields $N_D \gg$

N_A or $N_A \gg N_D$. Moreover, the intrinsic carrier concentration in Si at room temperature is about $10^{10}/cm^3$, while the dominant doping concentration (N_A or N_D) is seldom less than $10^{14}/cm^3$. Thus the special case considered here is the usual case encountered in practice. If $N_D - N_A \simeq N_D \gg n_i$, the square root in Eq. (2.29a) reduces to $N_D/2$ and

$$\boxed{\begin{array}{l} n \simeq N_D \\ p \simeq n_i^2/N_D \end{array}} \qquad \begin{array}{ll} N_D \gg N_A, \qquad N_D \gg n_i \\ \text{(nondegenerate, total ionization)} \end{array} \qquad \begin{array}{l} (2.30\text{a}) \\ (2.30\text{b}) \end{array}$$

Similarly

$$\boxed{\begin{array}{l} p \simeq N_A \\ n \simeq n_i^2/N_A \end{array}} \qquad \begin{array}{ll} N_A \gg N_D, \qquad N_A \gg n_i \\ \text{(nondegenerate, total ionization)} \end{array} \qquad \begin{array}{l} (2.31\text{a}) \\ (2.31\text{b}) \end{array}$$

As a numerical example, suppose a Si sample maintained at room temperature is uniformly doped with $N_D = 10^{15}/cm^3$ donors. Using Eqs. (2.30) one rapidly concludes $n \simeq 10^{15}/cm^3$ and $p \simeq 10^5/cm^3$.

(3) *Doped Semiconductor where* $n_i \gg |N_D - N_A|$. Systematically increasing the ambient temperature causes a monotonic rise in the intrinsic carrier concentration (see Fig. 2.20). At sufficiently elevated temperatures, n_i will eventually equal and then exceed the net doping concentration. If $n_i \gg |N_D - N_A|$, the square roots in Eqs. (2.29) reduce to n_i and $\boxed{n \simeq p \simeq n_i}$. In other words, *all semiconductors become intrinsic at sufficiently high temperatures where* $n_i \gg |N_D - N_A|$.

(4) *Compensated Semiconductor.* As is evident from Eqs. (2.29), donors and acceptors tend to negate each other. Indeed, it is possible to produce intrinsic-like material by making $N_D - N_A = 0$. In some materials, such as GaAs, N_A may be comparable to N_D in the as-grown crystal. When N_A and N_D are comparable and nonzero the material is said to be *compensated*. If the semiconductor is compensated, both N_A and N_D must be retained in all carrier concentration expressions.

In summary, Eqs. (2.29) can always be used to compute the carrier concentrations if the semiconductor is nondegenerate and the dopant sites are totally ionized. In the vast majority of practical situations, however, it is possible to simplify these equations prior to performing numerical computations. Equations (2.29) must be used to compute the carrier concentrations only in those rare instances when $|N_D - N_A| \sim n_i$. The simplified relationships of greatest practical utility are Eqs. (2.30) and (2.31).

SEE EXERCISE 2.3–APPENDIX A

2.5.6 Determination of E_F

Knowledge concerning the exact position of the Fermi level on the energy band diagram is often of interest. For example, when discussing the intrinsic Fermi level we indicated that E_i was located somewhere near the middle of the band gap. It would be useful to know the *precise* positioning of E_i in the band gap. Moreover, we have developed computational formulas for n and p appropriate for nondegenerate semiconductors. Whether a doped semiconductor is nondegenerate or degenerate depends, of course, on the value or positioning of E_F.

Before running through the mechanics of finding the Fermi level in selected cases of interest, it is useful to make a general observation; namely, Eqs. (2.19) or (2.16) [or even more generally, Eqs. (2.14)] provide a one-to-one correspondence between the Fermi energy and the carrier concentrations. Thus, having computed any one of the three variables — n, p, or E_F — one can always determine the remaining two variables under equilibrium conditions.

(1) Exact positioning of E_i. In an intrinsic material

$$n = p \tag{2.32}$$

Substituting for n and p in Eq. (2.32) using Eqs. (2.16), and setting $E_F = E_i$, yields

$$N_C e^{(E_i-E_C)/kT} = N_V e^{(E_V-E_i)/kT} \tag{2.33}$$

Solving for E_i, one obtains

$$E_i = \frac{E_c + E_v}{2} + \frac{kT}{2} \ln\left(\frac{N_V}{N_C}\right) \tag{2.34}$$

But

$$\frac{N_V}{N_C} = \left(\frac{m_p^*}{m_n^*}\right)^{3/2} \tag{2.35}$$

Consequently,

$$E_i = \frac{E_c + E_v}{2} + \frac{3}{4} kT \ln\left(\frac{m_p^*}{m_n^*}\right) \tag{2.36}$$

According to Eq. (2.36), E_i lies precisely at midgap only if $m_p^* = m_n^*$ or if $T = 0$ K. For the more practical case of silicon at room temperature, Table 2.1 gives $m_p^*/m_n^* = 0.69$, $(3/4)kT \ln(m_p^*/m_n^*) = -0.0073$ eV, and E_i therefore lies 0.0073 eV below midgap. Although potentially significant in certain problems, this small deviation from midgap is typically neglected in drawing energy band diagrams, etc.

(2) *Doped Semiconductors (nondegenerate, dopants totally ionized).* The general positioning of the Fermi level in donor- and acceptor-doped semiconductors assumed to be nondegenerate, in equilibrium, and maintained at temperatures where the dopants are fully ionized, is readily deduced from Eqs. (2.19). Specifically, solving Eqs. (2.19) for $E_F - E_i$, one obtains

$$E_F - E_i = kT \ln(n/n_i) = -kT \ln(p/n_i) \qquad (2.37)$$

Depending on the simplifications inherent in a particular problem, the appropriate carrier concentration solution [Eqs. (2.29), (2.30), (2.31), etc.] is then substituted into Eq. (2.37) to determine the positioning of E_F. For example, per Eqs. (2.30a) and (2.31a), $n \simeq N_D$ in typical donor-doped semiconductors and $p \simeq N_A$ in typical acceptor-doped semiconductors maintained at or near room temperature. Substituting into Eq. (2.37) we therefore conclude

$$
\boxed{
\begin{aligned}
E_F - E_i &= kT \ln(N_D/n_i) \\
E_i - E_F &= kT \ln(N_A/n_i)
\end{aligned}
}
\qquad
\begin{aligned}
&\ldots N_D \gg N_A, \quad N_D \gg n_i \qquad (2.38a)\\
&\ldots N_A \gg N_D, \quad N_A \gg n_i \qquad (2.38b)
\end{aligned}
$$

From Eqs. (2.38) it is obvious that the Fermi level moves systematically upward in energy from E_i with increasing donor doping and systematically downward in energy from E_i with increasing acceptor doping. The exact Fermi level positioning in Si at room temperature, nicely reinforcing the foregoing statement, is displayed in Fig. 2.21. Also note that for a given semiconductor material and ambient temperature there exist maximum nondegenerate donor and acceptor concentrations, doping concentrations above which the material becomes degenerate. In Si at room temperature the maximum nondegenerate doping concentrations are $N_D \simeq 1.6 \times 10^{18}/\text{cm}^3$ and $N_A \simeq 9.1 \times 10^{17}/\text{cm}^3$. The large Si doping values required for degeneracy, we should interject, have led to the common usage of "highly doped" (or n^+-material/p^+-material) and "degenerate" as interchangeable terms.

Finally, the question may arise, "What procedure should be followed in computing E_F when one is not sure whether a material is nondegenerate or degenerate?" Unless a material is known to be degenerate, always assume nondegeneracy and compute E_F employing the appropriate nondegenerate relationship. If E_F derived from the nonde-

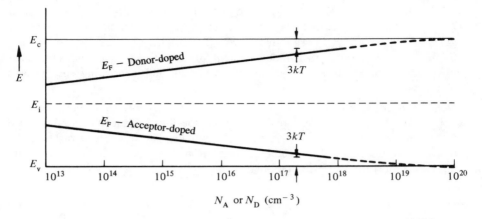

Fig. 2.21 Fermi level positioning in Si at room temperature as a function of the doping concentration. The solid E_F lines were established using Eq. (2.38a) for donor-doped material and Eq. (2.38b) for acceptor-doped material ($kT = 0.0259$ eV, and $n_i = 10^{10}/cm^3$).

generate formula lies in the degenerate zone, one must then, of course, recompute E_F using the more complex formalism valid for degenerate materials.

SEE EXERCISE 2.4–APPENDIX A

2.5.7 Carrier Concentration Temperature Dependence

A number of isolated facts about the carrier concentration temperature dependence have already been presented at various points in this chapter. The Section 2.3 discussion concerned with dopant action, for example, described the increased ionization of dopant sites and the associated increase in the majority carrier concentration when the temperature of a semiconductor is raised from near $T = 0$ K toward room temperature. More recent subsections have included a plot of the intrinsic carrier concentration versus temperature (Fig. 2.20) and a calculation indicating that all semiconductors become intrinsic ($n \rightarrow n_i$, $p \rightarrow n_i$) at sufficiently high temperatures. In this subsection, which concludes the carrier concentration discussion, temperature-related facts are combined and embellished to provide a broader, more complete description of how the carrier concentrations vary with temperature.

Figure 2.22(a), a typical majority-carrier concentration-versus-temperature plot constructed assuming a phosphorus-doped $N_D = 10^{15}/cm^3$ Si sample, nicely illustrates the general features of the concentration-versus-temperature dependence. Examining Fig. 2.22(a) we find that n is fixed at approximately N_D over a broad temperature range extending roughly from 150 K to 450 K for the given Si sample. This $n \simeq N_D$ or

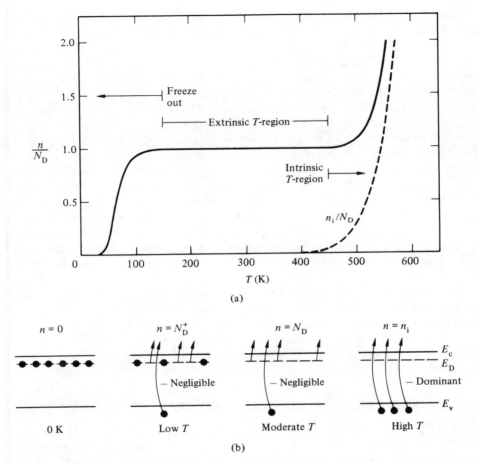

Fig. 2.22 (a) Typical temperature dependence of the majority-carrier concentration in a doped semiconductor. The plot was constructed assuming a phosphorus-doped $N_D = 10^{15}/\text{cm}^3$ Si sample. n_i/N_D versus T (dashed line) has been included for comparison purposes. (b) Qualitative explanation of the concentration-versus-temperature dependence displayed in part (a).

"extrinsic temperature region" constitutes the normal operating range for most solid state devices. Below 100 K or so, in the "freeze-out temperature region," n drops significantly below N_D and approaches zero as $T \to 0$ K. In the "intrinsic temperature region" at the opposite end of the temperature scale, n rises above N_D, asymptotically approaching n_i with increasing T.

To qualitatively explain the just-described concentration-versus-temperature dependence, it is important to recall that the equilibrium number of carriers within a material is affected by two separate mechanisms. Electrons donated to the conduction band from donor sites and valence band electrons excited across the band gap into the conduction band (broken Si–Si bonds) both contribute to the majority-carrier electron concentration in a donor-doped material. At temperatures $T \to 0$ K the thermal energy available in the system is insufficient to release the weakly bound fifth electron on donor sites and totally insufficient to excite electrons across the band gap. Hence $n = 0$ at $T = 0$ K, as visualized on the left-hand side of Fig. 2.22(b). Slightly increasing the material temperature above $T = 0$ K "defrosts" or frees some of the electrons weakly bound to donor sites. Band-to-band excitation, however, remains extremely unlikely, and therefore the number of observed electrons in the freeze-out temperature region equals the number of ionized donor sites — $n = N_D^+$. Continuing to increase the system temperature eventually frees almost all of the weakly bound electrons on donor sites, n approaches N_D, and one enters the extrinsic temperature region. In progressing through the extrinsic temperature region, more and more electrons are excited across the band gap, but the number of electrons supplied in this fashion stays comfortably below N_D. Ultimately, of course, electrons excited across the band gap equal, then exceed and, as pictured on the right-hand side of Fig. 2.22(b), finally swamp the fixed number of electrons derived from the donor sites.

As a practical note, it should be pointed out that the wider the band gap, the greater the energy required to excite electrons from the valence band into the conduction band, and the higher the temperature at the onset of the intrinsic temperature region. For this reason, Si devices can operate at higher temperatures than Ge devices, $\simeq 200°C$ at maximum for Si devices versus $\simeq 100°C$ at maximum for Ge devices. GaAs devices, of course, can operate at even higher temperatures than Si devices.

2.6 SUMMARY AND CONCLUDING COMMENTS

Under the general heading of carrier modeling we have described, examined, and characterized the carriers within a semiconductor under "rest" or equilibrium conditions. The many important topics addressed in this chapter included the introduction of two "visualization" models; namely, the bonding model and the energy band model. The extremely useful energy band model is actually more than just a model — it is a sophisticated sign language providing a concise means of communicating on a nonverbal level. Relative to the carriers themselves, the reader by now has been successfully prodded into thinking of electrons and holes as classical ball-like "particles," where the charge on an electron is $-q$, the charge on a hole is $+q$, and the effective masses of the particles are m_n^* and m_p^*, respectively. The reader should also know that the carrier numbers in an intrinsic material are equal and relatively small; the carrier concentrations, however, can be selectively increased by adding special impurity atoms or dopants to the semiconductor.

In addressing the problem of determining the carrier concentrations in doped semiconductors, we developed or derived a number of useful mathematical relationships. The density of states functions [Eqs. (2.6)], the Fermi function [Eq. (2.7)], the symmetrical nondegenerate relationships for n and p [Eqs. (2.19)], the np product [Eq. (2.22)], the charge neutrality relationship [Eq. (2.25)], and the simplified n and p expressions appropriate for a typical semiconductor maintained at room temperature [Eqs. (2.30) and (2.31)] deserve special mention. With regard to using the cited relationships, the reader should be cautioned against "no-think plug and chug." Because semiconductor problems are replete with exceptions, special cases, and nonideal situations, it is imperative that the formula user be aware of derivational assumptions and the validity limits of any and all expressions used in an analysis or computation. In addition to the quantitative carrier relationships, the reader should also have a qualitative "feel" for the carrier distributions in the respective energy bands, the temperature dependence of the intrinsic carrier concentration, and the typical temperature dependence of the majority carrier concentration in a doped semiconductor.

Finally, some attention should be given to the many technical terms and the key parametric values presented in this chapter. The terms extrinsic semiconductor, donor, acceptor, nondegenerate semiconductor, Fermi level, etc., will be encountered again and again in the discussion of semiconductor devices. Likewise, a knowledge of typical numerical values for key parameters, such as $E_G = 1.12$ eV and $n_i \cong 10^{10}/\text{cm}^3$ in Si at room temperature, will be invaluable in subsequent work when performing crude, order-of-magnitude computations. Key parametric values also serve as "yardsticks" for gauging whether newly encountered quantities are relatively small or relatively large.

PROBLEMS

2.1 Using the *bonding* model for a semiconductor, indicate how one visualizes: (a) a missing atom, (b) an electron, (c) a hole, (d) a donor, (e) an acceptor.

2.2 Using the *energy band* model for a semiconductor, indicate how one visualizes: (a) an electron, (b) a hole, (c) donor sites, (d) acceptor sites, (e) freeze-out of majority carrier electrons at donor sites as the temperature is lowered toward 0 K, (f) freeze-out of majority carrier holes at acceptor sites as the temperature is lowered toward 0 K, (g) the energy distribution of carriers in the respective bands, (h) an intrinsic semiconductor, (i) an n-type semiconductor, (j) a p-type semiconductor, (k) a nondegenerate semiconductor, (l) a degenerate semiconductor.

2.3 Energy-related questions.

(a) 1 eV is equal to how many joules of energy?

(b) kT is equal to how many eV at 300 K?

(c) At 300 K, $kT/q = ?$

(d) The ionization energy (equal to $|E_B|$) of phosphorus sites in Si is?

(e) $E_G(\text{Si}) = ?$

(f) $E_G(\text{GaAs}) = ?$

(g) The energy required to ionize a hydrogen atom initially in the $\mathbf{n} = 1$ state is?

(h) Why is the answer to part (g) significantly larger than the answer to part (d)?

2.4 (a) Under equilibrium conditions and $T > 0$ K, what is the probability of an electron state being occupied if it is located at the Fermi level?

(b) If E_F is positioned at E_c, determine (numerical answer required) the probability of finding electrons in states at $E_c + kT$.

2.5 The carrier distributions or numbers of carriers as a function of energy in the conduction and valence bands were noted to peak at an energy very close to the band edges. (See the carrier distribution sketches in Fig. 2.16.) Taking the semiconductor to be nondegenerate, determine the energy relative to the band edges at which the carrier distributions peak.

2.6 Starting with Eq. (2.8b) and following a procedure analogous to that outlined in the text, present the intermediate steps and arguments leading to Eqs. (2.14b) and (2.16b).

2.7 The density of states functions for a hypothetical semiconductor are:

$$g_c(E) = \text{constant} = N_C/kT \quad \ldots E \geq E_c$$

$$g_v(E) = \text{constant} = N_V/kT \quad \ldots E \leq E_v$$

Following the procedure outlined in the text, establish carrier concentration relationships for the hypothetical semiconductor analogous to Eqs. (2.14) and (2.16).

2.8 (a) Verify the text statement that, at 300 K,

$$N_{C,V} = (2.509 \times 10^{19}/\text{cm}^3)\,(m*/m_0)^{3/2}$$

where one sets $m* = m_n^*$ in computing N_C and $m* = m_p^*$ in computing N_V. $m_0 = 9.11 \times 10^{-31}$ kg; $h = 6.624 \times 10^{-34}$ joule-sec; and $q = 1.602 \times 10^{-19}$ coul.

(b) Using the effective mass values recorded in Table 2.1, construct a table which lists the numerical values of N_C and N_V for Si, Ge, and GaAs at 300 K.

2.9 Suppose m_n^* in Si was somehow made smaller than m_p^*. What effect would this reduction in m_n^* have on the density of states in the conduction band, N_C, n_i, and the position of E_i at room temperature?

2.10 Concentration questions with a reverse twist.

(a) A silicon wafer is uniformly doped p-type with $N_A = 10^{15}/\text{cm}^3$. At $T \simeq 0$ K, what are the equilibrium hole and electron concentrations?

(b) A semiconductor is doped with an impurity concentration N such that $N \gg n_i$ and all the impurities are ionized. Also, $n = N$ and $p = n_i^2/N$. Is the impurity a donor or an acceptor? Explain.

(c) The electron concentration in a piece of Si maintained at 300 K under equilibrium conditions is $10^5/\text{cm}^3$. What is the hole concentration?

(d) For a silicon sample maintained at $T = 300$ K the Fermi level is located 0.26 eV above the intrinsic Fermi level. What are the hole and electron concentrations?

2.11 Determine the equilibrium electron and hole concentrations inside a uniformly doped sample of Si under the following conditions:

(a) Room temperature, $N_A \ll N_D$, $N_D = 10^{15}/cm^3$;

(b) Room temperature, $N_A = 10^{16}/cm^3$, $N_D \ll N_A$;

(c) Room temperature, $N_A = 9 \times 10^{15}/cm^3$, $N_D = 10^{16}/cm^3$;

(d) $T = 450$ K, $N_A = 0$, $N_D = 10^{14}/cm^3$;

(e) $T = 650$ K, $N_A = 0$, $N_D = 10^{14}/cm^3$.

2.12 (a to e) For each of the conditions specified in Problem 2.11, determine the position of E_i, compute $E_F - E_i$, and draw a carefully dimensioned energy band diagram for the Si sample. Note: $E_G(Si) = 1.08$ eV at 450 K and 1.015 eV at 650 K.

2.13 According to the text, the maximum nondegenerate donor and acceptor doping concentrations in Si at room temperature are $N_D \simeq 1.6 \times 10^{18}/cm^3$ and $N_A \simeq 9.1 \times 10^{17}/cm^3$, respectively. Verify the text statement.

2.14 GaAs considerations.

(a) Make a sketch similar to Fig. 2.14 which is specifically appropriate for GaAs. Be sure to take into account the fact that $m_n^* \ll m_p^*$ in GaAs.

(b) Based on your answer to part (a), would you expect E_i in GaAs to lie above or below midgap? Explain.

(c) Determine the precise position of the intrinsic Fermi level in GaAs at room temperature (300 K).

(d) Determine the maximum nondegenerate donor and acceptor doping concentrations in GaAs at room temperature.

2.15 Given an $N_A = 10^{14}/cm^3$ doped Si sample:

(a) Present a *qualitative* argument which leads to the approximate positioning of the Fermi level in the material as $T \rightarrow 0$ K.

(b) Compute E_F as a function of T in the material at 50 K intervals from $T = 300$ K to $T = 500$ K.

(c) What do you conclude relative to the general behavior of the Fermi level positioning as a function of temperature?

(d) How would your foregoing answers be modified if the Si sample was doped with donors instead of acceptors?

3 / Carrier Action

Carrier modeling under "rest" or equilibrium conditions, considered in the previous chapter, is important because it establishes the proper frame of reference. From a device standpoint, however, the zero current observed under equilibrium conditions is singularly uninteresting. It is only when a semiconductor system is perturbed, giving rise to carrier action or a net carrier response, that currents can flow within and external to the semiconductor system. Action, carrier action, is the general concern of this chapter.

Under normal operating conditions the three primary types of carrier action occurring inside semiconductors are *drift, diffusion,* and *recombination–generation.* In this chapter we first describe each primary type of carrier action qualitatively and then quantitatively relate the action to the current flowing within the semiconductor. Special emphasis is placed on characterizing the "constants of the motion" associated with each type of action and, wherever appropriate, the discussion is extended to subsidiary topics of a relevant nature. Although introduced individually, the various types of carrier action are understood to occur simultaneously inside any given semiconductor. Putting everything together, so to speak, next leads to the culmination of our carrier-action efforts; we obtain the basic set of starting equations employed in solving device problems of an electrical nature. Finally, the chapter concludes with an examination of sample problems and problem solutions.

3.1 DRIFT

3.1.1 Definition–Visualization

Drift, by definition, *is charged-particle motion in response to an applied electric field.* Within semiconductors the drifting motion of the carriers on a microscopic scale can be described as follows: When an electric field ($\mathscr{E}$) is applied across a semiconductor as visualized in Fig. 3.1(a), the resulting force on the carriers tends to accelerate the $+q$ charged holes in the direction of the electric field and the $-q$ charged electrons in the direction opposite to the electric field. Because of collisions with ionized impurity atoms and thermally agitated lattice atoms, however, the carrier acceleration is frequently interrupted (the carriers are said to be scattered). The net result, pictured in

(a)   (b)   (c)

Fig. 3.1   Visualization of carrier drift: (a) motion of carriers within a biased semiconductor bar; (b) drifting hole on a microscopic or atomic scale; (c) carrier drift on a macroscopic scale.

Fig. 3.1(b), is carrier motion generally along the direction of the electric field, but in a disjointed fashion involving repeated periods of acceleration and subsequent decelerating collisions.

The microscopic drifting motion of a single carrier is obviously complex and quite tedious to analyze in any detail. Fortunately, measurable quantities are *macroscopic* observables which reflect the average or overall motion of the carriers. Averaging over all electrons or holes at any given time, we find that the resultant motion of each carrier type can be described in terms of a constant drift velocity, v_d. In other words, on a macroscopic scale, drift may be visualized [see Fig. 3.1(c)] as nothing more than all carriers of a given type moving along at a constant velocity in a direction parallel or antiparallel to the applied electric field.

By way of clarification, it is important to point out that the drifting motion of the carriers arising in response to an applied electric field is actually superimposed upon the always-present thermal motion of the carriers. Electrons in the conduction band and holes in the valence band gain and lose energy via collisions with the semiconductor lattice and are nowhere near stationary even under equilibrium conditions. In fact, under equilibrium conditions the thermally related carrier velocities average $\sim 1/1000$ the speed of light at room temperature! As pictured in Fig. 3.2, however, the thermal motion of the carriers is completely random. Thermal motion therefore averages out to zero on a macroscopic scale, does not contribute to current transport, and can be conceptually neglected.

Fig. 3.2   Thermal motion of a carrier.

3.1.2 Drift Current

Let us next turn to the task of developing an analytical expression for the current flowing within a semiconductor as a result of carrier drift. By definition

I (current) = the charge per unit time crossing an arbitrarily chosen plane of observation oriented normal to the direction of current flow.

Considering the p-type semiconductor bar of cross-sectional area A shown in Fig. 3.3, and specifically noting the arbitrarily chosen v_d-normal plane lying within the bar, we can argue:

$v_d t$ ...   All holes this distance back from the v_d-normal plane will cross the plane in a time t;

$v_d t A$ ...   All holes in this volume will cross the plane in a time t;

$p v_d t A$ ...   Holes crossing the plane in a time t;

$q p v_d t A$ ...   Charge crossing the plane in a time t;

$q p v_d A$ ...   Charge crossing the plane per unit time.

The word definition of the last quantity is clearly identical to the formal definition of current. Thus

$$I_{P|\text{drift}} = q p v_d A \qquad \text{hole drift current} \qquad (3.1)$$

As a practical matter, the cross-sectional area A appearing in Eq. (3.1) and other current formulas is often excess baggage. Current, moreover, is generally thought of as a scalar quantity, while in reality it is obviously a vector. These deficiencies are overcome by introducing a related parameter known as the current density, $\mathbf{J}$. $\mathbf{J}$ has the same orientation as the direction of current flow and is equal in magnitude to the current per unit area (or $J = I/A$). By inspection, the current density associated with hole drift is simply,

$$\mathbf{J}_{P|\text{drift}} = q p \mathbf{v}_d \qquad (3.2)$$

Fig. 3.3 Expanded view of a biased p-type semiconductor bar of cross-sectional area A.

Since the drift current arises in response to an applied electric field, it is reasonable to proceed one step further and seek a form of the current relationship which explicitly relates $J_{P|drift}$ to the applied electric field. To this end we note that, for small to moderate values of $\mathcal{E}$, the measured drift velocity in semiconductors (see Fig. 3.4) is directly proportional to the applied electric field. Theoretical analyses of a more fundamental nature also arrive at the same conclusion. Excluding situations involving large $\mathcal{E}$-fields, we can therefore write

$$v_d = \mu_p \mathcal{E} \tag{3.3}$$

where μ_p, the *hole mobility*, is the constant of proportionality between v_d and $\mathcal{E}$. Hence, substituting Eq. (3.3) into Eq. (3.2), one obtains

$$\boxed{J_{P|drift} = q\mu_p p \mathcal{E}} \tag{3.4a}$$

By a similar argument applied to electrons, one finds

$$\boxed{J_{N|drift} = q\mu_n n \mathcal{E}} \tag{3.4b}$$

where $J_{N|drift}$ is the electron current density due to drift and μ_n is the electron mobility.[†]

3.1.3 Mobility

Mobility is obviously a central parameter in characterizing electron and hole transport due to drift. As further readings will reveal, the carrier mobilities also play a key role in characterizing the performance of many devices. It is reasonable therefore to examine μ_n and μ_p in some detail to enhance our general familiarity with the parameters and to establish a core of useful information for future reference.

Standard Units: cm^2/V-sec.

Sample Numerical Values: $\mu_n \simeq 1360$ cm^2/V-sec and $\mu_p \simeq 460$ cm^2/V-sec at 300 K in $N_D = 10^{14}/cm^3$ and $N_A = 10^{14}/cm^3$ doped Si, respectively. In uncompensated high-purity (N_D or $N_A \leq 10^{15}/cm^3$) GaAs, the room-temperature mobilities are projected to

---

[†]Although electrons drift in the direction opposite to the applied electric field ($v_d = -\mu_n\mathcal{E}$), the current transported by negatively charged particles is in turn counter to the direction of drift ($J_N = -qnv_d$). The net result, as indicated in Eq. (3.4b), is an electron current in the direction of the applied electric field. It is also important to note, relative to Eqs. (3.3) and (3.4), that μ_n and μ_p are defined to be positive constants.

Fig. 3.4 Measured drift velocity of the carriers in ultrapure silicon maintained at room temperature as a function of the applied electric field.

be $\mu_n \simeq 8000$ cm^2/V-sec and $\mu_p \simeq 320$ cm^2/V-sec. The quoted values are useful for comparison purposes and when performing order-of-magnitude computations. Also note that $\mu_n > \mu_p$ for both Si and GaAs. In the major semiconductors, μ_n is consistently greater than μ_p for a given doping and system temperature.

Relationship to Scattering: Mobility is a measure of the ease of carrier motion within a semiconductor crystal. Increasing the motion-impeding collisions within a crystal decreases the mobility of the carriers. In other words, the carrier mobility varies inversely with the amount of scattering taking place within the semiconductor. As visualized in Fig. 3.1(b), the dominant scattering mechanisms in nondegenerately doped materials of device quality are typically (i) lattice scattering involving collisions with thermally agitated lattice atoms, and (ii) ionized impurity (i.e., donor-site and/or acceptor-site) scattering.

Doping Dependence: Figure 3.5 exhibits the observed doping dependence of the electron and hole mobilities in silicon at room temperature. The figure is also typical of the doping dependence observed in other semiconductor materials. At low doping concentrations, below approximately 10^{15}/cm^3 in Si, the carrier mobilities are essentially independent of the doping concentration. For dopings in excess of $\sim 10^{15}$/cm^3, the mobilities monotonically decrease with increasing N_A or N_D.

The explanation of the observed doping dependence is relatively straightforward. At sufficiently low doping levels, ionized impurity scattering can be neglected compared to lattice scattering. When lattice scattering, which is not a function of N_A or N_D, becomes the dominant scattering mechanism, it automatically follows that the carrier mobilities will be likewise independent of N_A or N_D. For dopings in excess of $\sim 10^{15}$/cm^3,

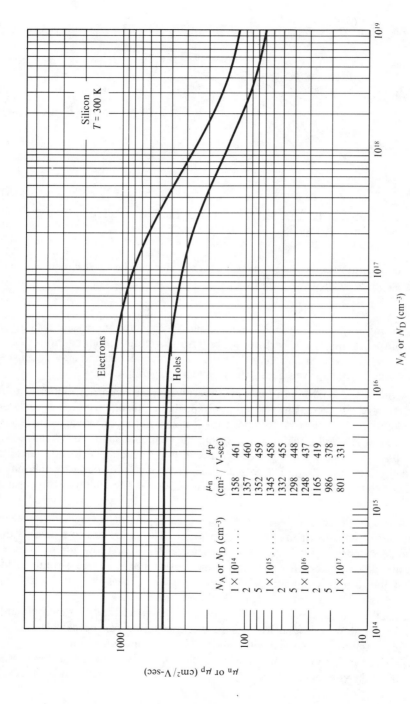

The following data appears within the figure:

N_A or N_D (cm^{-3})	μ_n (cm^2/V-sec)	μ_p (cm^2/V-sec)
1×10^{14}	1358	461
2	1357	460
5	1352	459
1×10^{15}	1345	458
2	1332	455
5	1298	448
1×10^{16}	1248	437
2	1165	419
5	986	378
1×10^{17}	801	331

Silicon
$T = 300$ K

Electrons

Holes

N_A or N_D (cm^{-3})

μ_n or μ_p (cm^2/V-sec)

Fig. 3.5 Room-temperature carrier mobilities in silicon as a function of the dopant concentration. μ_n is the electron mobility; μ_p is the hole mobility.

ionized impurity scattering can no longer be neglected. Increasing the number of scattering centers by adding more and more acceptors or donors in excess of $10^{15}/cm^3$ clearly increases the amount of nonnegligible ionized impurity scattering and correspondingly decreases the carrier mobilities.

Temperature Dependence: The temperature dependence of the electron and hole mobilities in Si with doping as a parameter is displayed in Fig. 3.6. In the lowest-doped samples the carrier mobilities monotonically decrease as the temperature is increased, where roughly $\mu_n \propto T^{-2.3\pm0.1}$ and $\mu_p \propto T^{-2.2\pm0.1}$. For higher sample dopings the carrier mobilities systematically decrease with doping at a given temperature and are far less temperature sensitive.

The general temperature dependence of the carrier mobilities in the lightly doped samples is relatively easy to explain. Increasing the system temperature causes an ever-increasing thermal agitation of the semiconductor atoms, which in turn increases the lattice scattering. Since lattice scattering is the dominant scattering mechanism in the lightly doped samples, increasing the system temperature therefore monotonically decreases the mobility of the carriers.

For higher sample dopings the temperature dependence reflects the added effects of ionized impurity scattering. Ionized impurity scattering is not only significant in the more heavily doped samples, but may even become the dominant scattering mechanism at lower temperatures. This is true because, unlike lattice scattering, ionized impurity scattering exhibits an inverse temperature dependence, *decreasing* as the temperature is increased. Although we will omit the details, an appropriate combination of the lattice scattering component, which increases with temperature, and the ionized impurity scattering component, which decreases with temperature, can account for any of the curves presented in Fig. 3.6.

3.1.4 Resistivity

Resistivity is an important material parameter which is closely related to carrier drift. Qualitatively, resistivity is a measure of a material's inherent resistance to current flow — a "normalized" resistance that does not depend on the physical dimensions of the material. Quantitatively, resistivity (ρ) is defined as the proportionality constant between the electric field impressed across a homogeneous material and the total particle current per unit area flowing in the material; that is,

$$\mathscr{E} = \rho \mathbf{J} \tag{3.5a}$$

or

$$\mathbf{J} = \sigma \mathscr{E} = \frac{1}{\rho}\mathscr{E} \tag{3.5b}$$

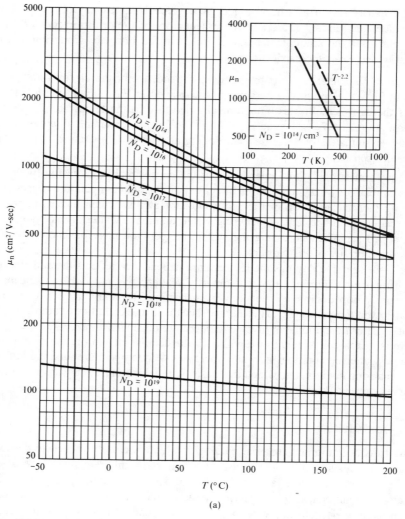

Fig. 3.6 Temperature dependence of (a) electron and (b) hole mobilities in silicon for sample dopings ranging from $10^{14}/cm^3$ to $10^{19}/cm^3$. In the lightest-doped sample, $\mu \propto T(K)^{-n}$. This is clearly evident from the inset plots of log μ versus log $T(K)$.

where $\sigma = 1/\rho$ is the material conductivity. In a homogeneous material, $\mathbf{J} = \mathbf{J}_{drift}$ and, as established with the aid of Eqs. (3.4),

$$\mathbf{J}_{drift} = \mathbf{J}_{N|drift} + \mathbf{J}_{P|drift} = q(\mu_n n + \mu_p p)\mathscr{E} \qquad (3.6)$$

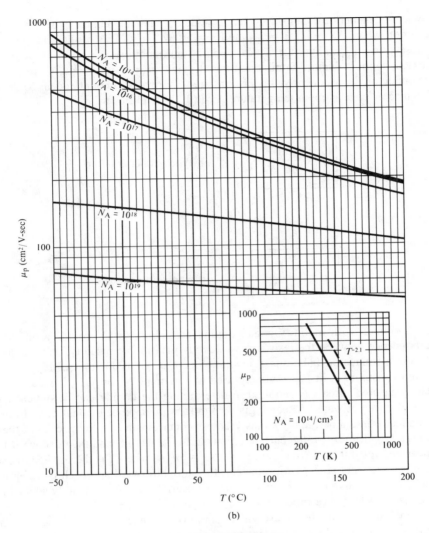

Fig. 3.6   (continued)

It therefore follows that

$$\rho = \frac{1}{q(\mu_n n + \mu_p p)}$$

(3.7)

In a nondegenerate donor-doped semiconductor maintained in the extrinsic tempera-ture region where $N_D \gg n_i$, $n \simeq N_D$, and $p \simeq n_i^2/N_D \ll n$. This result was established in Subsection 2.5.5. Thus, for typical dopings and mobilities, $\mu_n n + \mu_p p \simeq \mu_n N_D$ in an n-type semiconductor. Similar arguments yield $\mu_n n + \mu_p p \simeq \mu_p N_A$ in a p-type semiconductor. Consequently, under conditions normally encountered in Si samples maintained at or near room temperature, Eq. (3.7) simplifies to

$$\boxed{\rho = \frac{1}{q\mu_n N_D}} \quad \ldots n\text{-type semiconductor} \tag{3.8a}$$

and

$$\boxed{\rho = \frac{1}{q\mu_p N_A}} \quad \ldots p\text{-type semiconductor} \tag{3.8b}$$

When combined with mobility-versus-doping data, Eqs. (3.8) provide a one-to-one correspondence between the resistivity, a directly measurable quantity, and the doping inside a semiconductor. In conjunction with plots of ρ versus doping (see Fig. 3.7), the measured resistivity is in fact routinely used to determine N_A or N_D.

The measured resistivity required in determining the doping can be obtained in a number of different ways. A seemingly straightforward approach would be to form the semiconductor into a bar, apply a bias V across contacts attached to the ends of the bar as in Fig. 3.1(a), measure the current I flowing in the circuit, and then deduce ρ from the measured resistance. [R(resistance) $= V/I = \rho l/A$, where l is the bar length and A is the cross-sectional area.] Unfortunately, the straightforward approach is deceptively difficult, is destructive (wastes semiconductor material), and is not readily adaptable to the wafers used in device processing.

The measurement method most widely employed in practice is the four-point probe technique. In the standard four-point probe technique, four collinear, evenly spaced probes, as shown in Fig. 3.8, are brought into contact with the surface of the semicon-ductor. A known current I is passed through the outer two probes and the potential V thereby developed is measured across the inner two probes. The semiconductor resis-tivity is then computed from

$$\rho = 2\pi s \frac{V}{I} \mathcal{F} \tag{3.9}$$

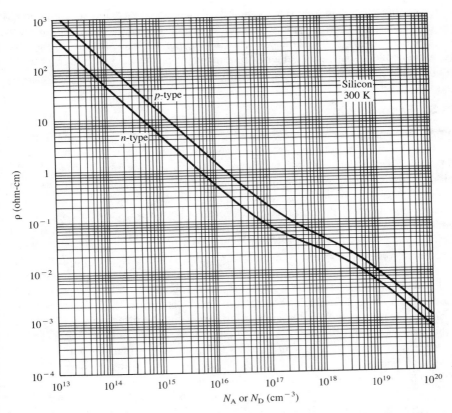

Fig. 3.7 Si resistivity versus impurity concentration at 300 K.

Fig. 3.8 Schematic drawing of the probe arrangement, placement, and biasing in the four-point probe measurement.

where s is the probe-to-probe spacing and $\mathscr{F}$ is a tabulated or graphed "correction" factor. The correction factor typically depends on the thickness of the sample and on whether the bottom of the semiconductor is touching an insulator or a metal. Unlike the semiconductor-bar measurement, the four-point probe technique is obviously easy to implement, nondestructive, and designed for use in the processing of wafers.[†]

3.1.5 Band Bending

In previous encounters with the energy band diagram we have consistently drawn E_c and E_v to be energies independent of the position coordinate x. When an electric field ($\mathscr{E}$) exists inside a material, the band energies become a function of position. The resulting variation of E_c and E_v with position on the energy band diagram, exemplified by Fig. 3.9(a), is popularly referred to as "band bending."

Seeking to establish the precise relationship between the electric field within a semiconductor and the induced band bending, let us carefully re-examine the energy band diagram. The diagram itself, as emphasized in Fig. 3.9(a), is a plot of the allowed electron energies within the semiconductor as a function of position, where E increasing upward is understood to be the *total* energy of the electrons. Furthermore, we know from previous discussions that if an energy of precisely E_G is added to break a semiconductor-semiconductor bond, the created electron and hole energies would be E_c and E_v, respectively, and the created carriers would be effectively motionless. Absorbing an energy in excess of E_G, on the other hand, would in all probability give rise to an electron energy greater than E_c and a hole energy less than E_v, with both carriers moving around rapidly within the lattice. We are led, therefore, to interpret $E - E_c$ to be the kinetic energy (K. E.) of the electrons and $E_v - E$ to be the kinetic energy of the holes [see Fig. 3.9(b)]. Moreover, since the total energy equals the sum of the kinetic energy and the potential energy (P. E.), E_c minus the energy reference level (E_{ref}) must equal the electron potential energy, as illustrated in Fig. 3.9(c). (Potential energy, it should be noted, is arbitrary to within a constant, and the position-independent reference energy, E_{ref}, may be chosen to be any convenient value.)

The potential energy is the key in relating the electric field within a semiconductor to positional variations in the energy bands. Specifically, assuming normal operational conditions, where magnetic field, temperature gradient, and stress-induced effects are negligible, only the force associated with an existing electric field can give rise to changes in the potential energy of the carriers. Elementary physics, in fact, tells us that the potential energy of a $-q$ charged particle under such conditions is simply related to the electrostatic potential V at a given point by

$$\text{P. E.} = -qV \tag{3.10}$$

[†]For an excellent and detailed discussion of the four-point probe technique, a full tabulation of correction factors, and an examination of other in-use resistivity measurements, the reader is referred to Chapter 3 in W. R. Runyan, *Semiconductor Measurements and Instrumentation*, McGraw-Hill, New York, 1975.

Fig. 3.9 Relationship between band bending and the electrostatic variables inside a semiconductor: (a) sample energy band diagram exhibiting band bending; (b) identification of the carrier kinetic energies; (c) specification of the electron potential energy; (d) electrostatic potential and (e) electric field versus position dependence deduced from and associated with the part (a) energy band diagram.

Having previously concluded that

$$\text{P. E.} = E_c - E_{ref} \tag{3.11}$$

we can state

$$\boxed{V = -\frac{1}{q}(E_c - E_{ref})} \tag{3.12}$$

By definition, moreover,

$$\mathscr{E} = -\nabla V \tag{3.13}$$

or, in one dimension,

$$\mathscr{E} = -\frac{dV}{dx} \tag{3.14}$$

Consequently,

$$\boxed{\mathscr{E} = \frac{1}{q}\frac{dE_c}{dx} = \frac{1}{q}\frac{dE_v}{dx} = \frac{1}{q}\frac{dE_i}{dx}} \tag{3.15}$$

The latter forms of Eq. (3.15) follow from the fact that E_c, E_v and E_i differ by only an additive constant.

The preceding formulation provides a means of readily deducing the general form of the electrostatic variables associated with the "band bending" in Fig. 3.9(a) and other energy band diagrams. Making use of Eq. (3.12), or simply inverting $E_c(x)$ in Fig. 3.9(a), one obtains the V versus x dependence presented in Fig. 3.9(d). [Please note that the electrostatic potential, like the potential energy, is arbitrary to within a constant — the Fig. 3.9(d) plot can be translated upward or downward along the voltage axis without modifying the physical situation inside the semiconductor.] Finally, recording the slope of E_c versus position, as dictated by Eq. (3.15), produces the $\mathscr{E}$ versus x plot shown in Fig. 3.9(e).

In summary, the reader should be aware of the fact that the energy band diagram contains information relating to the electrostatic potential and electric field within the semiconductor. Moreover, the general form of the V and $\mathscr{E}$ dependencies within the semiconductor can be obtained almost by inspection. To deduce the form of the V versus x relationship, merely sketch the "upsidedown" of E_c (or E_v or E_i) versus x; to determine the general $\mathscr{E}$ versus x dependence, simply note the slope of E_c (or E_v or E_i) as a function of position.

```
SEE EXERCISE 3.1(a–d) – APPENDIX A
```

3.2 DIFFUSION

3.2.1 Definition–Visualization

Diffusion is a process whereby particles tend to spread out or redistribute as a result of their random thermal motion, migrating on a macroscopic scale from regions of high particle concentration into regions of low particle concentration. If allowed to progress unabated, the diffusion process operates so as to produce a uniform distribution of particles. The diffusing entity, it should be noted, need not be charged; thermal motion, not interparticle repulsion, is the enabling action behind the diffusion process.

To cite an everyday example of diffusion, suppose an open bottle of perfume is placed in one corner of a room. Even in the absence of air currents, random thermal motion will spread the perfume molecules throughout the room in a relatively short period of time, with intermolecular collisions helping to uniformly redistribute the perfume to every nook and cranny within the room.

Seeking to obtain a more detailed understanding of the diffusion process, let us next conceptually "monitor" the process on a microscopic scale employing a simple hypothetical system. The system we propose to monitor is a one-dimensional box containing four compartments and 1024 mobile particles (see Fig. 3.10). The particles within the box obey certain stringent rules. Specifically, thermal motion causes all particles in a given compartment to "jump" into an adjacent compartment every τ_0 seconds. In keeping with the random nature of the motion, each and every particle has an equal probability of jumping to the left and to the right. Hitting an "external wall" while attempting to jump to the left or right reflects the particle back to its pre-jump position. Finally, at time $t = 0$ it is assumed that all of the particles are confined in the left-most compartment.

Figure 3.10 records the evolution of our 1024 particle system as a function of time. At time $t = \tau_0$, 512 of the 1024 particles originally in compartment 1 jump to the right and come to rest in compartment 2. The remaining 512 particles jump to the left and are reflected back into the left-most compartment. The end result is 512 particles each in compartments 1 and 2 after τ_0 seconds. At time $t = 2\tau_0$, 256 of the particles in compartment 2 jump into compartment 3, while the remainder jump back into compartment 1. In the meantime, 256 of the particles from compartment 1 jump into compartment 2 and 256 undergo a reflection at the left-hand wall. The net result after $2\tau_0$ seconds is 512 particles in compartment 1 and 256 particles each in compartments 2 and 3. The state of the system after $3\tau_0$ and $6\tau_0$ seconds, also shown in Fig. 3.10, can be deduced in a similar manner. By $t = 6\tau_0$, the particles, once confined to the left-most compartment, have become almost uniformly distributed throughout the box, and it is unnecessary to consider later states. Indeed, the fundamental nature of the diffusion process is clearly self-evident from an examination of the existing states.

Fig. 3.10 Diffusion on a microscopic scale in a hypothetical one-dimensional system. The numbers over the arrows indicate the number of particles in a given compartment; observation times are listed to the extreme right.

In semiconductors the diffusion process on a microscopic scale is similar to that occurring in the hypothetical system except, of course, the random motion of the diffusing particles is three-dimensional and not "compartmentalized." On a *macroscopic* scale the net effect of diffusion is precisely the same within both the hypothetical system and semiconductors; there is an overall migration of particles from regions of high particle concentration to regions of low particle concentration. Within semiconductors the mobile particles, the electrons and holes, are charged, and diffusion-related carrier transport therefore gives rise to particle currents as pictured in Fig. 3.11.

3.2.2 Hot-Point Probe Measurement

Before establishing the current associated with diffusion, let us digress somewhat and briefly consider the hot-point probe measurement. The hot-point probe measurement is a common technique for rapidly determining whether a semiconductor is *n*- or *p*-type.

Fig. 3.11   Visualization of electron and hole diffusion on a macroscopic scale.

From a practical standpoint, knowledge of the semiconductor type is essential in device processing and must be known even before determining the doping concentration from resistivity measurements (refer back to Fig. 3.7). The hot-point probe "typing" experiment is considered here because it simultaneously provides an informative example of the diffusion process.

Examining Fig. 3.12(a), we find the only equipment required for performing the hot-point probe measurement is a hot-point probe, a cold-point probe, and a center-zero milliammeter. The hot-point probe is sometimes a refugee from a wood burning set or simply a soldering iron; the cold-point probe typically assumes the form of an electrical probe like that used with hand-held multimeters. No special requirements are imposed on the center-zero milliammeter connected between the probes. The measurement procedure itself is also extremely simple: After allowing the hot probe to heat up, one brings the two probes into contact with the semiconductor sample, and the ammeter deflects to the right or left, thereby indicating the semiconductor type. Observe that the spacing between the probes is set arbitrarily and can be reduced to enhance the ammeter deflection.

A simplified explanation of how the hot-point probe measurement works is presented in Fig. 3.12(b). In the vicinity of the probe contact the heated probe creates an increased number of higher-energy carriers. These energetic carriers will be predominantly holes in the case of a p-type material and electrons in an n-type material. With

Fig. 3.12   The hot-point probe measurement: (a) required equipment; (b) simplified explanation of how the measurement works.

more energetic carriers near the heated probe than elsewhere, diffusion acts so as to spread the higher-energy carriers throughout the semiconductor wafer. The net effect is a deficit of holes or a net negative charge in the vicinity of the hot-point probe for a p-type material, and a positive charge buildup near the heated probe within an n-type material. Accordingly, the center-zero meter deflects in a different direction for p- and n-type materials.

3.2.3 Diffusion Currents/Total Carrier Currents

In defining the diffusion process and in citing diffusion examples we have sought to emphasize the direct correlation between diffusion and a spatial variation in particle numbers. In order for diffusion to occur, more of the diffusing particles must exist at one point than at other points or, in mathematical terms, there must be a nonzero concentration gradient ($\nabla p \neq 0$ for holes, $\nabla n \neq 0$ for electrons). Logically, moreover, the greater the concentration gradient, the larger the expected flux of carriers and the larger the associated carrier currents. Analysis of the diffusion process indeed confirms the preceding expectations. Currents arising from diffusion are, in fact, directly proportional to the particle concentration gradients. Explicitly, for electrons and holes,

$$\mathbf{J}_{P|diff} = -qD_P\nabla p \tag{3.16a}$$

$$\mathbf{J}_{N|diff} = qD_N\nabla n \tag{3.16b}$$

where the constants of proportionality, D_P and D_N, have units of cm^2/sec and are referred to as the hole and electron diffusion coefficients, respectively.

Upon examining Eqs. (3.16), please note that the current directions deduced from the equations are consistent with the Fig. 3.11 visualization of the macroscopic diffusion currents. For the positive concentration gradient shown in Fig. 3.11 ($dp/dx > 0$ and $dn/dx > 0$ for the pictured one-dimensional situation), both holes and electrons will diffuse in the $-x$ direction. $J_{P|diff}$ will therefore be negative or directed in the $-x$ direction while $J_{N|diff}$ will be oriented in the $+x$ direction.

With explicit expressions available for both the drift and diffusion components of the carrier currents flowing inside of a semiconductor, it is now a trivial matter to establish expressions for the total carrier currents themselves. Simply combining the respective n and p segments of Eqs. (3.4) and Eqs. (3.16), one obtains

$$\mathbf{J}_P = q\mu_p p\mathscr{E} - qD_P\nabla p \tag{3.17a}$$
$$\updownarrow \text{ drift } \quad \updownarrow \text{ diffusion}$$
$$\mathbf{J}_N = q\mu_n n\mathscr{E} + qD_N\nabla n \tag{3.17b}$$

The double box around Eqs. (3.17) emphasizes the importance of the total-current relationships, which are used directly or indirectly in essentially all device analyses.

3.2.4 Einstein Relationship

The diffusion coefficients, the constants of the motion associated with diffusion, are obviously central parameters in characterizing carrier transport due to diffusion. Given the importance of the diffusion coefficients, one might anticipate an extended examination of relevant properties paralleling the mobility presentation in Subsection 3.1.3. Fortunately, an extended examination is unnecessary because the D's and the μ's are interrelated. It is only necessary to establish the connecting formula known as the Einstein relationship.

Before performing the mathematical manipulations leading to the Einstein relationship, let us examine a nondegenerate, *nonuniformly* doped semiconductor maintained under *equilibrium conditions*. A concrete example of what we have in mind is shown in Fig. 3.13. Figure 3.13(a) exhibits the assumed doping variation with position and Fig. 3.13(b) the corresponding equilibrium energy band diagram. Two facts inherent in the diagram are absolutely essential in establishing the Einstein relationship. (These facts are also important in themselves.) First of all, it is a fundamental law of physics that

> *under equilibrium conditions the Fermi level inside a material* (or inside a group of materials in intimate contact) *is invariant as a function of position;* that is, $dE_F/dx = dE_F/dy = dE_F/dz = 0$ under equilibrium conditions.

As shown in Fig. 3.13(b), E_F is at the same energy for all x. The Fermi energy is analogous to the water level in a swimming pool or any other container. Regardless of the contour assumed by the bottom of the pool or container, the water line is always

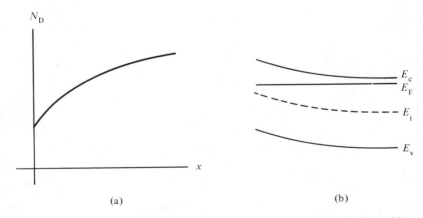

(a)   (b)

Fig. 3.13 Nonuniformly doped semiconductor: (a) assumed doping variation with position; (b) corresponding equilibrium energy band diagram.

level if the system is unperturbed. Secondly, in Chapter 2 we found that the Fermi level in uniformly doped n-type semiconductors moved closer and closer to E_c when the donor doping was systematically increased (see Fig. 2.21). Consistent with this fact and as diagramed in Fig. 3.13(b), the distance between E_c and E_F is smaller in regions of higher doping. Band bending is therefore a natural consequence of spatial variations in doping, and

> *a nonzero electric field is established inside nonuniformly doped semiconductors under equilibrium conditions.*

With the preliminary considerations completed, we can now proceed to the derivation proper. Under equilibrium conditions the total carrier currents inside a semiconductor must be identically zero. Still considering a nondegenerate, nonuniformly doped semiconductor under equilibrium conditions, and simplifying the presentation by working in only one dimension, we can therefore state

$$J_{N|\text{drift}} + J_{N|\text{diff}} = q\mu_n n\mathscr{E} + qD_N\frac{dn}{dx} = 0 \tag{3.18}$$

However,

$$\mathscr{E} = \frac{1}{q}\frac{dE_i}{dx} \tag{3.19}$$

$$\text{(same as 3.15)}$$

and

$$n = n_i e^{(E_F - E_i)/kT} \tag{3.20}$$

Moreover, with $dE_F/dx = 0$ under equilibrium conditions,

$$\frac{dn}{dx} = -\frac{n_i}{kT}e^{(E_F-E_i)/kT}\frac{dE_i}{dx} = -\frac{q}{kT}n\mathscr{E} \tag{3.21}$$

Substituting dn/dx from Eq. (3.21) into Eq. (3.18), and rearranging the result slightly, one obtains

$$(qn\mathscr{E})\mu_n - (qn\mathscr{E})\frac{q}{kT}D_N = 0 \tag{3.22}$$

Since $\mathscr{E} \neq 0$ (a consequence of the nonuniform doping), it follows from Eq. (3.22) that

$$\boxed{\frac{D_N}{\mu_n} = \frac{kT}{q}} \qquad \text{Einstein relationship for electrons} \tag{3.23a}$$

A similar argument for holes yields

$$\boxed{\frac{D_P}{\mu_p} = \frac{kT}{q}}$$ Einstein relationship for holes   (3.23b)

Although it was established while assuming equilibrium conditions, we can present more elaborate arguments which show the Einstein relationship to be valid even under nonequilibrium conditions. The nondegenerate restriction, however, still applies; slightly modified forms of Eqs. (3.23) result when the argument is extended to degenerate materials. We should emphasize again that the intermediate results concerning the positional invariance of the equilibrium Fermi level, and the existence of a nonzero electric field inside a nonuniformly doped semiconductor under equilibrium conditions, are important in themselves. Relative to numerical values, note that kT/q is a voltage, and at room temperature is approximately equal to 0.026 V.[†] Hence, for a $N_D = 10^{14}/cm^3$ Si sample maintained at room temperature, $D_N = (kT/q)\mu_n \cong (0.026 \text{ V})(1360 \text{ cm}^2/\text{V-sec}) \cong 35 \text{ cm}^2/\text{sec}$. Finally, the Einstein relationship is one of the easiest equations to remember because it rhymes internally—Dee over mu equals kTee over q. The rhyme holds even if the equation is inverted—mu over Dee equals q over kTee.

SEE EXERCISE 3.1(e–h)–APPENDIX A

3.3 RECOMBINATION–GENERATION

3.3.1 Definition–Visualization

Recombination–generation is perhaps the most interesting of the three primary types of carrier action and at the same time the most challenging to totally master. Formally, recombination–generation can be defined as follows:

Generation is a process whereby electrons and holes (carriers) are created.

Recombination is a process whereby electrons and holes are annihilated or destroyed.

Unlike drift and diffusion, generation and the inverse action of recombination are not single processes, but instead are collective names for a group of similar processes; that is, there are a number of different ways in which carriers can be created and destroyed within a semiconductor. The simplest of the R–G processes are visualized in Fig. 3.14. Figure 3.14(a) models the photogeneration process where photons (light) with an energy greater than E_G impinge upon the semiconductor and create electrons and holes by

---

[†]At room temperature $kT \cong 0.026$ eV $= (1.6 \times 10^{-19})(0.026)$ joule. Thus $kT/q = (1.6 \times 10^{-19})(0.026)/(1.6 \times 10^{-19})$ joule/coul $= 0.026$ V.

(a) Photogeneration

(b) Direct thermal generation

(c) Direct thermal recombination

Fig. 3.14 Band-to-band recombination–generation processes. Bonding model (left-hand side) and equivalent energy band model (right-hand side) visualizations of (a) photogeneration, (b) direct thermal generation, and (c) direct thermal recombination.

breaking semiconductor–semiconductor bonds. Alternatively, the photogeneration process may be visualized in terms of the energy band diagram as light energy being absorbed and exciting electrons from the valence band into the conduction band. The process shown in Fig. 3.14(b) is essentially identical to the Fig. 3.14(a) process, except the excitation is due to thermal (or heat) energy present within the material. Note that the Fig. 3.14(b) action is referred to as *direct* thermal generation. Finally, Fig. 3.14(c) pictures *direct* thermal recombination, the exact inverse of the Fig. 3.14(b) process. An electron and hole moving in the semiconductor lattice, and generally minding their own business, stray into the same spatial vicinity and zap! — the electron and hole annihilate each other.

If a semiconductor happens to be illuminated, photogeneration will typically occur precisely as pictured in Fig. 3.14(a). It is a fact of life, however, that although direct band-to-band recombination–generation may be simple to visualize and easy to characterize, the Fig. 3.14(b) and (c) processes are *not* the usual means whereby carriers are thermally created and annihilated in the major semiconductors. The thermal creation and annihilation of carriers, an action going on at all times in all semiconductors, is typically dominated by *indirect* thermal recombination–generation.

Indirect thermal recombination–generation, pictured in Fig. 3.15, involves a "third party" or intermediary and takes place only at special locations within the semiconductor known as R–G centers. Physically, R–G centers are special impurity atoms (such as gold, iron, and copper in Si) or lattice defects [such as the missing atom in Fig. 3.15(a)]. Lattice defects and the special atoms in the form of unintentional impurities are present even in the major semiconductors, although the R–G center concentration in device-quality materials is normally small compared to the acceptor and donor concentrations. The most important property of the R–G centers is depicted in Fig. 3.15(b); namely, R–G centers (or traps,[†] for short) introduce allowed electronic levels near the center of the band gap. The near midgap positioning of the E_T levels is all-important because it distinguishes R–G centers from donors and acceptors.

Having cited pertinent R–G center details, let us next briefly examine how carriers are created and destroyed by interacting with the centers. In Fig. 3.15(a) we envision an electron wandering through the semiconductor lattice until it strays into the vicinity of a lattice defect and is caught in the potential well associated with the defect. A little later in time a hole comes along, is attracted to the negatively charged electron, and proceeds to annihilate the electron (and itself) at the R–G center site. The energy band version of precisely the same process is shown on the left-hand side of Fig. 3.15(b). An electron travels to the vicinity of an R–G center, loses energy, and is trapped. Subsequently, a hole wanders into the vicinity of the same center, loses energy, and annihilates the electron within the center. (Equivalently, one can think of the electron as jumping down a second time and annihilating the hole in the valence band. We should also note that it is possible for the hole to be captured first.) The inverse process, carrier generation, is shown on the right-hand side of Fig. 3.15(b). After absorbing thermal energy, an electron first jumps up to the E_T level, creating a hole in the

(a)   (b)

Fig. 3.15 R–G center (indirect thermal) recombination–generation. (a) Bonding model visualization of recombination at a missing atom R–G center. (b) Energy band model description of the R–G center assisted recombination and generation of carriers. R–G centers introduce the E_T levels pictured near the center of the band gap.

[†]To the device specialist the terms "R–G center" and "trap" refer to the same type of physical center but describe somewhat different modes of operation. The subscript T, nevertheless, is almost universally employed to identify R–G center parameters such as the E_T energy level in Fig. 3.15(b).

valence band. The creation process is completed when the electron absorbs additional thermal energy and jumps into the conduction band (hence becoming a carrier).

Having examined the basic R–G models and information, we are finally in a position to address the inevitable question of how recombination–generation fits into the overall scheme of things. First of all, it must be understood that the thermal recombination and generation of carriers are never-ending processes. Under equilibrium conditions the recombination and generation rates balance and there is no net change in the carrier numbers. On the other hand, when the semiconductor is perturbed, the thermal recombination and generation rates automatically change so as to favor a return to equilibrium conditions. If the perturbation leads to a carrier excess, the thermal recombination rate becomes greater than the generation rate, thereby favoring a reduction in carrier numbers. If the perturbation causes a carrier deficit, generation will occur at a faster rate than recombination. In other words, thermal R–G is nature's mechanism for stabilizing the carrier concentrations within a material and for reinstating the "most preferred" condition of equilibrium. Photogeneration, we should caution, is a separate process and occurs in addition to thermal R–G when the semiconductor is illuminated. Photogeneration always acts so as to create an excess of carriers. From the foregoing statements, then, we see that the major effect of the R–G action is to produce a change in the carrier concentrations. Recombination–generation indirectly affects current flow in semiconductors by manipulating the carrier concentrations involved in the drift and diffusion processes.

3.3.2   R–G Statistics

R–G statistics is the technical name given to the mathematical characterization of recombination–generation processes. "Mathematical characterization" in the case of recombination–generation does not mean the development of a current density relationship; the R–G action does not lead *per se* to current transport. Instead, as was pointed out in the last subsection, recombination–generation changes the carrier concentrations, and it is the time rate of change in the carrier concentrations ($\partial n/\partial t$, $\partial p/\partial t$) that must be specified to achieve a mathematical characterization of any R–G process.

To illustrate the general formalism, let us initially consider the simple case of photogeneration. The photogeneration process gives rise to an equal number of added holes and electrons per second, and thus one can write

$$\left. \frac{\partial n}{\partial t} \right|_{\text{light}} = \left. \frac{\partial p}{\partial t} \right|_{\text{light}} = G_{\text{L}} \qquad (\text{number/cm}^3\text{-sec}) \qquad (3.24)$$

The $|_{\text{light}}$ designation means "due to light" and is necessary because n and p can change via other processes. Greater carrier drift or diffusion into a region than out of a region or vice versa, and other forms of recombination–generation, can also change the carrier concentrations. G_{L}, it should be noted, will in general be a function of the distance the light has penetrated into the material and the wavelength of the light. To achieve closed-form solutions or for illustrative purposes, however, G_L is often taken to be a constant, and the illumination is then said to be uniform.

Thermal R–G statistics, and in particular the statistics for the typically dominant R–G center mechanism, are far more complex than the photogeneration statistics. The general case development is, in fact, beyond the scope of this introductory volume. It is possible, nevertheless, to present plausible arguments leading to thermal R–G rate expressions which are valid under circumstances encountered again and again in practical problems.

Let us begin the special case development by defining:

n_0, p_0	carrier concentrations in the material under analysis when equilibrium conditions prevail.
n, p	carrier concentrations in the material under arbitrary conditions.
$\Delta n = n - n_0;$ $\Delta p = p - p_0$	deviations in the carrier concentrations from their equilibrium values. Δn and Δp can be both positive and negative, where a positive deviation corresponds to a carrier excess and a negative deviation corresponds to a carrier deficit.
$\left.\dfrac{\partial n}{\partial t}\right\|_{\substack{\text{thermal}\\ \text{R–G}}};$ $\left.\dfrac{\partial p}{\partial t}\right\|_{\substack{\text{thermal}\\ \text{R–G}}}$	time rate of change in the carrier concentrations due to *both* the indirect thermal recombination and indirect thermal generation processes occurring at R–G centers.
N_T	number of R–G centers/cm^3.

We next require that the semiconductor be (1) decidedly n- or p-type and (2) subject to a perturbation which causes only *low level injection*. In Si the first requirement is seldom violated. The second requirement is just an elegant way of saying the perturbation must be relatively small. To be more precise,

---

low level injection implies

$\Delta p \ll n_0,$ $n \simeq n_0$ in an n-type material;

$\Delta n \ll p_0,$ $p \simeq p_0$ in a p-type material.

Consider a specific example of $N_D = 10^{14}/\text{cm}^3$ doped Si at room temperature subject to a perturbation where $\Delta p = \Delta n = 10^9/\text{cm}^3$. For the given material, $n_0 \simeq N_D = 10^{14}/\text{cm}^3$ and $p_0 \simeq n_i^2/N_D \simeq 10^6/\text{cm}^3$. Consequently, $n = n_0 + \Delta n \simeq n_0$ and $\Delta p = 10^9/\text{cm}^3 \ll n_0 = 10^{14}/\text{cm}^3$. The situation is clearly one of low level injection. Observe, however, that $\Delta p \gg p_0$. Although the majority carrier concentration remains essentially unperturbed under low level injection, the minority carrier concentration can, and routinely does, increase by many orders of magnitude.

Having disposed of the preliminaries, let us consider the situation shown in Fig. 3.16. The semiconductor is clearly n-type and a perturbation has caused a

Fig. 3.16   Situation inside an n-type semiconductor after a perturbation causing the low level injection of excess holes.

$\Delta p \ll n_0$ excess of holes. Assuming the system is in the midst of relaxing back to the equilibrium state from the pictured perturbed state via the R–G center interaction, what factors would you expect to have the greatest effect on the rate of relaxation (i.e., on $\partial p / \partial t |_{\text{thermal R–G}}$)? Well, to get rid of an excess hole, the hole must make the transition from the valence band to an electron-filled R–G center. Logically, the greater the number of filled R–G centers, the greater the probability of a hole annihilating transition and the faster the rate of relaxation. Since essentially all of the R–G center levels are filled with electrons because $E_F > E_T$ in our n-type semiconductor, and since the multitude of electrons rapidly fill any levels that become vacant, the number of filled R–G centers during the relaxation process is $\simeq N_T$. Thus, we expect $\partial p / \partial t |_{\text{thermal R–G}}$ to be approximately proportional to N_T. Moreover, it also seems logical that the number of hole annihilating transitions should increase almost linearly with the number of excess holes. All other factors being equal, the more holes available for annihilation, the greater the number of holes recombining per second. Consequently, we also expect $\partial p / \partial t |_{\text{thermal R–G}}$ to be approximately proportional to Δp. Although other factors might be considered, there are no additional dependencies to be established. Hence, introducing a positive proportionality constant, c_p, and realizing that $\partial p / \partial t |_{\text{thermal R–G}}$ decreases (is negative) when $\Delta p > 0$, we conclude

$$\left. \frac{\partial p}{\partial t} \right|_{\substack{\text{thermal} \\ \text{R–G}}} = - c_p N_T \Delta p \qquad \text{for holes in an } n\text{-type material} \qquad (3.25a)$$

and by a similar argument,

$$\left.\frac{\partial n}{\partial t}\right|_{\substack{\text{thermal}\\ \text{R-G}}} = -c_n N_T \Delta n \qquad \text{for electrons in a } p\text{-type material} \qquad (3.25\text{b})$$

where c_n is a positive proportionality constant distinct from c_p. c_n and c_p are referred to as the electron and hole capture coefficients, respectively.

If one examines the left-hand sides of Eqs. (3.25), the dimensional units are found to be those of a concentration divided by time. Since Δp and Δn on the right-hand sides of the same equations are also concentrations, the constants $c_p N_T$ and $c_n N_T$ must have units of 1/time. It is therefore reasonable to introduce the time constants

$$\tau_p = \frac{1}{c_p N_T} \qquad (3.26\text{a})$$

$$\tau_n = \frac{1}{c_n N_T} \qquad (3.26\text{b})$$

which, when substituted into Eqs. (3.25), yield

$$\left.\frac{\partial p}{\partial t}\right|_{\substack{\text{thermal}\\ \text{R-G}}} = -\frac{\Delta p}{\tau_p} \qquad \text{for holes in an } n\text{-type material} \qquad (3.27\text{a})$$

$$\left.\frac{\partial n}{\partial t}\right|_{\substack{\text{thermal}\\ \text{R-G}}} = -\frac{\Delta n}{\tau_n} \qquad \text{for electrons in a } p\text{-type material} \qquad (3.27\text{b})$$

Equations (3.27) are the desired end result, the special case characterization of R–G center (indirect thermal) recombination–generation. Steady state or quasi steady state (meaning "nearly" steady state) conditions are implicitly assumed in the development of Eqs. (3.27). In practice, however, the relationships can be applied with little error to most transient problems of interest. As indicated previously, the relationships also only apply to minority carriers and to situations meeting the low level injection requirement.[†] Finally, please note that a $\Delta p < 0$ will give rise to a $\partial p/\partial t|_{\text{thermal R-G}} > 0$. A positive

---

[†]Provided steady state conditions prevail and $E_T = E_i$, it can be established (see Vol. VI, Chapter 5) that for arbitrary injection levels and any nondegenerate doping

$$\left.\frac{\partial p}{\partial t}\right|_{\substack{\text{thermal}\\ \text{R-G}}} = \left.\frac{\partial n}{\partial t}\right|_{\substack{\text{thermal}\\ \text{R-G}}} = \frac{np - n_i^2}{\tau_p(n + n_i) + \tau_n(p + n_i)}$$

The above general relationship is included for completeness and future reference. The reader should verify that the general relationship reduces to Eqs. (3.27) under the special conditions (low level injection, etc.) assumed herein.

$\partial p/\partial t\,|_{\text{thermal R–G}}$ simply indicates that a carrier deficit exists inside the semiconductor and generation is occurring at a more rapid rate than recombination. $\partial p/\partial t\,|_{\text{thermal R–G}}$ and $\partial n/\partial t\,|_{\text{thermal R–G}}$, it must be remembered, characterize the *net* effect of the thermal recombination and thermal generation processes.

3.3.3 Minority Carrier Lifetimes

The time constants τ_n and τ_p, which were introduced without comment in writing down Eqs. (3.26), are obviously the "constants of the action" associated with recombination–generation. Seeking to provide insight relative to the standard interpretation and naming of the τ's, let us consider once again the situation pictured in Fig. 3.16. Examining the change in the hole concentration with time, it goes almost without saying that the excess holes do not all disappear at the same time. Rather, the hole excess present at $t = 0$ is systematically eliminated, with some of the holes existing for only a short period and others "living" for comparatively long periods of time. If thermal recombination–generation is the sole process acting to relax the semiconductor, the average excess hole lifetime, $\langle t \rangle$, can be computed in a relatively straightforward manner. Without going into details, the computation yields $\langle t \rangle = \tau_n$ (or τ_p). Physically, therefore, τ_n and τ_p have come to be interpreted as *the average time an excess minority carrier will live in a sea of majority carriers*. For identification purposes, τ_n and τ_p are simply referred to as the minority carrier lifetimes.

Given the importance of the τ_n and τ_p parameters in the modeling of devices, one might expect at this point to encounter a sizeable list of relevant facts and information about the minority carrier lifetimes. Actually, a catalog of well-defined properties similar to those of the carrier mobilities just does not exist. In fact, subsidiary experimental measurements must be performed to determine the minority carrier lifetime in a given semiconductor sample. The reason for this apparent lack of information can be traced to the extreme variability of the τ_n and τ_p parameters. Referring to Eqs. (3.26), we see that the carrier lifetimes depend on the often poorly controlled R–G center concentration and *not* on the carefully controlled doping parameters (N_A and N_D). Moreover, the physical nature of the dominant R–G center can vary from sample to sample, and the R–G center concentration changes even within a given sample during device fabrication. A fabrication procedure called "gettering" can reduce the R–G center concentration to a very low level and give rise to a $\tau_n(\tau_p) \sim 1$ msec in Si. The intentional introduction of gold into Si, on the other hand, can controllably increase the R–G center concentration and give rise to a $\tau_n(\tau_p) \sim 1$ nsec. Typical minority carrier lifetimes in most Si devices lie somewhere between the cited extremes; usually $\tau_n(\tau_p)$ are $\sim 1\ \mu$sec.

3.4 EQUATIONS OF STATE

In the first three sections of this chapter we examined and separately modeled the primary types of carrier action taking place inside semiconductors. Within actual semicon-

ductors the various types of carrier action all occur at the same time, and the state of
any semiconductor system can be determined only by taking into account the combined
effect of the individual types of carrier action. "Putting it all together," so to speak,
leads to the basic set of starting equations employed in solving device problems, herein
referred to as the *equations of state*. This section is devoted to developing the equations of state and to surveying common simplifications and special-case solutions.

3.4.1 Continuity Equations

Each and every type of carrier action, whether it be drift, diffusion, thermal recombination, thermal generation, or some other type of carrier action, gives rise to a change
in the carrier concentrations with time. The combined effect of all types of carrier action
can therefore be taken into account by equating the overall change in the carrier con-
centrations per unit time ($\partial n/\partial t$ or $\partial p/\partial t$) to the sum of the $\partial n/\partial t$'s or $\partial p/\partial t$'s due to
the individual processes; that is,

$$\frac{\partial n}{\partial t} = \frac{\partial n}{\partial t}\bigg|_{\text{drift}} + \frac{\partial n}{\partial t}\bigg|_{\text{diff}} + \frac{\partial n}{\partial t}\bigg|_{\substack{\text{thermal}\\\text{R-G}}} + \frac{\partial n}{\partial t}\bigg|_{\substack{\text{other processes}\\\text{(photogen., etc.)}}} \tag{3.28a}$$

$$\frac{\partial p}{\partial t} = \frac{\partial p}{\partial t}\bigg|_{\text{drift}} + \frac{\partial p}{\partial t}\bigg|_{\text{diff}} + \frac{\partial p}{\partial t}\bigg|_{\substack{\text{thermal}\\\text{R-G}}} + \frac{\partial p}{\partial t}\bigg|_{\substack{\text{other processes}\\\text{(photogen., etc.)}}} \tag{3.28b}$$

The overall effect of the individual processes is established, in essence, by invoking
the requirement of conservation of carriers. Electrons and holes cannot mysteriously
appear and disappear at a given point, but must be transported to or created at the
given point via some type of ongoing carrier action. There must be a spatial and time
continuity in the carrier concentrations. For this reason Eqs. (3.28) are known as the
continuity equations.

The continuity equations can be written in a somewhat more compact form by noting

$$\frac{\partial n}{\partial t}\bigg|_{\text{drift}} + \frac{\partial n}{\partial t}\bigg|_{\text{diff}} = \frac{1}{q}\left(\frac{\partial J_{\text{N}x}}{\partial x} + \frac{\partial J_{\text{N}y}}{\partial y} + \frac{\partial J_{\text{N}z}}{\partial z}\right) = \frac{1}{q}\nabla\cdot\mathbf{J}_{\text{N}} \tag{3.29a}$$

$$\frac{\partial p}{\partial t}\bigg|_{\text{drift}} + \frac{\partial p}{\partial t}\bigg|_{\text{diff}} = -\frac{1}{q}\left(\frac{\partial J_{\text{P}x}}{\partial x} + \frac{\partial J_{\text{P}y}}{\partial y} + \frac{\partial J_{\text{P}z}}{\partial z}\right) = -\frac{1}{q}\nabla\cdot\mathbf{J}_{\text{P}} \tag{3.29b}$$

Equations (3.29), which can be established by a straightforward mathematical manipulation, merely state that there will be a change in the carrier concentrations within a
given small region of the semiconductor if an imbalance exists between the total carrier
currents into and out of the region. Utilizing Eqs. (3.29), we obtain

$$\frac{\partial n}{\partial t} = \frac{1}{q} \nabla \cdot \mathbf{J_N} + \frac{\partial n}{\partial t}\bigg|_{\substack{\text{thermal} \\ \text{R-G}}} + \frac{\partial n}{\partial t}\bigg|_{\substack{\text{other} \\ \text{processes}}} \qquad (3.30\text{a})$$

$$\frac{\partial p}{\partial t} = -\frac{1}{q} \nabla \cdot \mathbf{J_P} + \frac{\partial p}{\partial t}\bigg|_{\substack{\text{thermal} \\ \text{R-G}}} + \frac{\partial p}{\partial t}\bigg|_{\substack{\text{other} \\ \text{processes}}} \qquad (3.30\text{b})$$

The (3.30) continuity equations are completely general and directly or indirectly constitute the starting point in most device analyses. In computer simulations the continuity equations are often employed directly. The appropriate relationships for $\partial n/\partial t\,|_{\text{thermal R-G}}$, $\partial p/\partial t\,|_{\text{thermal R-G}}$ [which may or may not be the special-case relationships given by Eqs. (3.27)], along with the concentration changes due to "other processes," are substituted into Eqs. (3.30), and numerical solutions are sought for $n(x, y, z, t)$ and $p(x, y, z, t)$. In problems where a closed-form type of solution is desired, the continuity equations are typically utilized only in an indirect fashion. The actual starting point in such analyses is a simplified version of the continuity equations to be established in the next subsection.

3.4.2 Minority Carrier Diffusion Equations

The "workhorse" minority carrier diffusion equations are derived from the continuity equations by invoking the following set of simplifying assumptions:

1. The particular system under analysis is *one-dimensional;* i.e., all variables are at most a function of just one coordinate (say the x-coordinate).

2. The analysis is limited or restricted to *minority carriers.*

3. $\mathscr{E} \simeq 0$ in the semiconductor or regions of the semiconductor subject to analysis.

4. The equilibrium minority carrier concentrations are not a function of position. In other words, $n_0 \neq n_0(x)$, $p_0 \neq p_0(x)$.

5. *Low level injection* conditions prevail.

6. There are *no "other processes,"* except possibly photogeneration, taking place within the system.

Working on the continuity equation for electrons, we note that

$$\frac{1}{q} \nabla \cdot \mathbf{J_N} \rightarrow \frac{1}{q} \frac{\partial J_N}{\partial x} \qquad (3.31)$$

if the system is one-dimensional (simplification #1). Moreover,

$$J_N = q\mu_n n \mathscr{E} + q D_N \frac{\partial n}{\partial x} \simeq q D_N \frac{\partial n}{\partial x} \tag{3.32}$$

when $\mathscr{E} \simeq 0$ and one is concerned only with minority carriers (simplifications #2 and #3). By way of explanation, the drift component can be neglected in the current density expression because $\mathscr{E}$ is small by assumption and minority carrier concentrations are also small, making the $n\mathscr{E}$ product extremely small. [Note that the same argument cannot be applied to majority carriers.] Since, by assumption (simplification #4), $n_0 \neq n_0(x)$, and by definition $n = n_0 + \Delta n$, we can also write

$$\frac{\partial n}{\partial x} = \frac{\partial n_0}{\partial x} + \frac{\partial \Delta n}{\partial x} = \frac{\partial \Delta n}{\partial x} \tag{3.33}$$

Combining Eqs. (3.31) through (3.33) yields

$$\frac{1}{q} \nabla \cdot \mathbf{J_N} \rightarrow D_N \frac{\partial^2 \Delta n}{\partial x^2} \tag{3.34}$$

Turning to the remaining terms in the continuity equation for electrons, the low level injection restriction (simplification #5) combined with the minority carrier limitation (simplification #2) permits us to employ the special-case expression for $\partial n/\partial t\,|_{\text{thermal R–G}}$ derived in Subsection 3.3.2.

$$\left.\frac{\partial n}{\partial t}\right|_{\substack{\text{thermal} \\ \text{R–G}}} = -\frac{\Delta n}{\tau_n} \tag{3.35}$$

In addition, simplification #6 allows us to write

$$\left.\frac{\partial n}{\partial t}\right|_{\substack{\text{other} \\ \text{processes}}} = G_L \tag{3.36}$$

where it is understood that $G_L = 0$ if the semiconductor is not subject to illumination. Finally, the equilibrium electron concentration is never a function of time, $n_0 \neq n_0(t)$, and we can therefore write

$$\frac{\partial n}{\partial t} = \frac{\partial n_0}{\partial t} + \frac{\partial \Delta n}{\partial t} = \frac{\partial \Delta n}{\partial t} \tag{3.37}$$

Substituting Eqs. (3.34) through (3.37) into the (3.30a) continuity equation, and simultaneously recording the analogous result for holes, one obtains

$$\boxed{\begin{aligned} \frac{\partial \Delta n_p}{\partial t} &= D_N \frac{\partial^2 \Delta n_p}{\partial x^2} - \frac{\Delta n_p}{\tau_n} + G_L \\ \frac{\partial \Delta p_n}{\partial t} &= D_P \frac{\partial^2 \Delta p_n}{\partial x^2} - \frac{\Delta p_n}{\tau_p} + G_L \end{aligned}}$$

Minority carrier
diffusion equations

(3.38a)

(3.38b)

We have added subscripts to the carrier concentrations in Eqs. (3.38) to remind the user that the equations are valid only for minority carriers, applying to electrons in p-type materials and to holes in n-type materials.

3.4.3 Simplifications and Solutions

In the course of performing device analyses the conditions of a problem often permit additional simplifications that drastically reduce the complexity of the minority carrier diffusion equations. More often than not, however, a simplification is stated in words, and difficulty is sometimes encountered in perceiving the mathematical consequences of the simplification. To assist the reader, we have tabulated below the most commonly encountered simplification statements and the corresponding mathematical modifications of the minority carrier diffusion equations. We also note below solutions to forms of the minority carrier diffusion equations resulting from certain combinations of the cited simplifications, solutions one is likely to encounter in later readings.

Simplification statement	Mathematical simplification
Steady state	$\dfrac{\partial \Delta n_p}{\partial t} \to 0 \quad \left(\dfrac{\partial \Delta p_n}{\partial t} \to 0\right)$
No concentration gradient or no diffusion current	$D_N \dfrac{\partial^2 \Delta n_p}{\partial x^2} \to 0 \quad \left(D_P \dfrac{\partial^2 \Delta p_n}{\partial x^2} \to 0\right)$
No drift current or $\mathscr{E} = 0$	No further simplification. ($\mathscr{E} \simeq 0$ is assumed in the derivation.)
No thermal R–G	$\dfrac{\Delta n_p}{\tau_n} \to 0 \quad \left(\dfrac{\Delta p_n}{\tau_p} \to 0\right)$
No light	$G_L \to 0$

Solution no. 1

GIVEN: Steady state, no light.

SIMPLIFIED
DIFF. EQN:
$$0 = D_N \frac{d^2 \Delta n_p}{dx^2} - \frac{\Delta n_p}{\tau_n}$$

SOLUTION: $\Delta n_p(x) = A e^{-x/L_N} + B e^{x/L_N}$

        where $L_N \equiv \sqrt{D_N \tau_n}$

        and A, B are solution constants.

Solution no. 2

GIVEN: No concentration gradient, no light.

SIMPLIFIED
DIFF. EQN:
$$\frac{d \Delta n_p}{dt} = -\frac{\Delta n_p}{\tau_n}$$

SOLUTION: $\Delta n_p(t) = \Delta n_p(0) e^{-t/\tau_n}$

Solution no. 3

GIVEN: Steady state, no concentration gradient.

SIMPLIFIED
DIFF. EQN:
$$0 = -\frac{\Delta n_p}{\tau_n} + G_L$$

SOLUTION: $\Delta n_p = G_L \tau_n$

Solution no. 4

GIVEN: Steady state, no R−G, no light.

SIMPLIFIED
DIFF. EQN:
$$0 = D_N \frac{d^2 \Delta n_p}{dx^2} \quad \text{or} \quad 0 = \frac{d^2 \Delta n_p}{dx^2}$$

SOLUTION: $\Delta n_p(x) = A + Bx$

3.4.4   Equations Summary

For the reader's convenience we have collected in this subsection carrier action and related equations routinely encountered in device analyses. Except for (3.41) and (3.44), the equations are repetitions of previously established relationships. The reader is assumed to be familiar with Eq. (3.41), which is a well-known expression from the field of electricity and magnetism. Equation (3.44), on the other hand, simply states that the total current density under steady state conditions (**J**) is the sum of the electron and hole current densities. Please note that the symbol ρ appearing in Eqs. (3.41) and (3.43) is the charge density, the charge/cm^3, and not the resistivity. It is unfortunate that the same symbol has come to be closely identified with two different quantities, but this duality never leads to difficulties because the meaning of the symbol can always be inferred from the context of the problem. Also, in writing down the continuity equations we have introduced simplified symbols for the various $\partial n / \partial t$'s and $\partial p / \partial t$'s. The simplified continuity equation symbols and all other newly introduced symbols are defined at the end of the subsection.

$$\frac{\partial n}{\partial t} = \frac{1}{q} \nabla \cdot \mathbf{J}_N - r_N + g_N \qquad \left(\text{Continuity} \atop \text{equations} \right) \qquad (3.39a)$$

$$\frac{\partial p}{\partial t} = -\frac{1}{q} \nabla \cdot \mathbf{J}_P - r_P + g_P \qquad\qquad\qquad (3.39b)$$

$$\frac{\partial \Delta n_p}{\partial t} = D_N \frac{\partial^2 \Delta n_p}{\partial x^2} - \frac{\Delta n_p}{\tau_n} + G_L \qquad \left(\text{Minority carrier} \atop \text{diffusion equations} \right) \qquad (3.40a)$$

$$\frac{\partial \Delta p_n}{\partial t} = D_P \frac{\partial^2 \Delta p_n}{\partial x^2} - \frac{\Delta p_n}{\tau_p} + G_L \qquad\qquad\qquad (3.40b)$$

$$\nabla \cdot \mathscr{E} = \rho / K_S \varepsilon_0 \qquad \text{(Poisson's equation)} \qquad (3.41)$$

where

$$\mathbf{J}_N = q \mu_n n \mathscr{E} + q D_N \nabla n \qquad (3.42a)$$

$$\mathbf{J}_P = q \mu_p p \mathscr{E} - q D_P \nabla p \qquad (3.42b)$$

$$\rho = q(p - n + N_D^+ - N_A^-) \qquad (3.43)$$

$$\mathbf{J} = \mathbf{J}_N + \mathbf{J}_P \qquad (3.44)$$

and

$$r_N \equiv -\frac{\partial n}{\partial t} \bigg|_{\substack{\text{thermal} \\ \text{R-G}}}; \qquad r_P \equiv -\frac{\partial p}{\partial t} \bigg|_{\substack{\text{thermal} \\ \text{R-G}}};$$

$$g_N \equiv \frac{\partial n}{\partial t} \bigg|_{\substack{\text{other} \\ \text{processes}}}; \qquad g_P \equiv \frac{\partial p}{\partial t} \bigg|_{\substack{\text{other} \\ \text{processes}}};$$

ρ = charge density, charge/cm^3;

K_S = semiconductor dielectric constant;

ε_0 = permittivity of free space;

$\mathbf{J}$ = total current density under steady state conditions.

3.5 PROBLEM SOLVING AND SUPPLEMENTAL CONCEPTS

A person encountering the topic of carrier action for the first time is apt to be overwhelmed. How does one use all these formulas? Where does one start in solving actual problems? Device analyses are, of course, examples of problem solutions, and most of the reader's questions will be answered in subsequent discussions. Nevertheless, it is worthwhile to examine some simple problems to indicate suggested problem-solving approaches. Moreover, the sample problems reviewed herein serve the additional purpose of introducing two supplemental concepts, minority carrier diffusion lengths and quasi-Fermi levels, concepts which are encountered in a number of device analyses.

3.5.1 Sample Problem No. 1

Problem: A uniformly donor-doped silicon wafer maintained at room temperature is suddenly illuminated with light at time $t = 0$. Assuming $N_D = 10^{15}/\text{cm}^3$, $\tau_p = 10^{-6}$ sec, and a light-induced creation of 10^{17} electrons and holes per cm^3-sec throughout the semiconductor, determine $\Delta p_n(t)$ for $t > 0$.

Solution: *Step 1* — Review precisely what information is given or implied in the statement of the problem.

The semiconductor is silicon, T = room temperature, the donor doping is the same everywhere with $N_D = 10^{15}/\text{cm}^3$, and $G_L = 10^{17}/\text{cm}^3$-sec at all points inside the semiconductor. Also, the statement of the problem *implies* equilibrium conditions exist for $t < 0$.

Step 2 — Characterize the system under equilibrium conditions.

For Si at room temperature $n_i \cong 10^{10}/\text{cm}^3$. Since $N_D \gg n_i$, $n_0 = N_D = 10^{15}/\text{cm}^3$ and $p_0 = n_i^2/N_D = 10^5/\text{cm}^3$. Because of the uniform doping, the equilibrium n_0 and p_0 values are, of course, the same everywhere throughout the semiconductor.

Step 3 — Analyze the problem qualitatively.

Prior to $t = 0$, equilibrium conditions prevail and $\Delta p_n = 0$. Starting at $t = 0$ the light creates added electrons and holes and Δp_n will begin to increase. The growing excess carrier numbers, however, in turn lead to an enhanced thermal recombination rate, with $\partial p/\partial t\,|_{\text{thermal R-G}} = -\Delta p_n/\tau_p$. Consequently, as Δp_n increases as a result of photogeneration, more and more of the excess holes are eliminated per second by thermal recombination through R–G centers. Eventually, a point is reached where the carriers annihilated per second by thermal recombination balance the carriers created per second by the light, and a steady state condition is attained.

Summarizing, we expect $\Delta p_n(t)$ to start from zero at $t = 0$, to build up at a decreasing rate, and to ultimately become constant. Since $\partial p/\partial t\,|_{\text{light}} + \partial p/\partial t\,|_{\text{thermal R-G}} = 0$ or $G_L - \Delta p_n/\tau_p = 0$ under steady state conditions, we can even state that $\Delta p_n(t \to \infty) = G_L\tau_p$.

Step 4 — Perform a quantitative analysis.

The minority carrier diffusion equation is the starting point for most first-order quantitative analyses. After examining the problem for obvious conditions which would invalidate the use of the diffusion equation, the appropriate minority carrier diffusion equation is written down, the equation is simplified, and a solution is sought subject to boundary conditions stated or implied in the problem.

For the problem under consideration a cursory inspection reveals that all simplifying assumptions involved in deriving the diffusion equations are readily satisfied. Specifically, only the minority carrier concentration is of interest; the equilibrium carrier concentrations are not a function of position, and there are no "other processes" except for photogeneration. Because the photogeneration is uniform throughout the semiconductor, the perturbed carrier concentrations are also position-independent

($\Delta p_n \neq \Delta p_n(x, y, z)$) and the electric field $\mathscr{E}$ must clearly be zero in the perturbed system. Finally, since $\Delta p_{n|max} = G_L \tau_p = 10^{11}/cm^3 \ll n_0 = 10^{15}/cm^3$, low level injection conditions prevail at all times.

With no obstacles to utilizing the diffusion equation, the desired quantitative solution can now be obtained by solving

$$\frac{\partial \Delta p_n}{\partial t} = D_P \frac{\partial^2 \Delta p_n}{\partial x^2} - \frac{\Delta p_n}{\tau_p} + G_L \qquad (3.45)$$

subject to the boundary condition

$$\Delta p_n(t)|_{t=0} = 0 \qquad (3.46)$$

Since $\Delta p_n \neq \Delta p_n(x)$, the diffusion equation simplifies to

$$\frac{d \Delta p_n}{dt} + \frac{\Delta p_n}{\tau_p} = G_L \qquad (3.47)$$

The general solution of Eq. (3.47) is

$$\Delta p_n(t) = G_L \tau_p + A e^{-t/\tau_p} \qquad (3.48)$$

Applying the boundary condition yields

$$A = -G_L \tau_p \qquad (3.49)$$

and

$$\Delta p_n(t) = G_L \tau_p (1 - e^{-t/\tau_p}) \quad \Leftarrow \text{solution} \qquad (3.50)$$

Step 5 — Examine the solution.

Failing to examine the mathematical solution to a problem is like growing vegetables and then failing to eat the produce. Relative to the Eq. (3.50) result, $G_L \tau_p$ has the dimensions of a concentration (number/cm^3) and the solution is at least dimensionally correct. A plot of the Eq. (3.50) result is shown in Fig. 3.17. Note that, in agreement with qualitative predictions, $\Delta p_n(t)$ starts from zero at $t = 0$ and eventually saturates at $G_L \tau_p$ after a few τ_p.

3.5.2   Sample Problem No. 2

Problem: As pictured in Fig. 3.18(a), the $x = 0$ end of a uniformly doped semi-infinite bar of silicon with $N_D = 10^{15}/cm^3$ is illuminated so as to create $\Delta p_{n0} = 10^{10}/cm^3$ excess holes at $x = 0$. The wavelength of the illumination is such that no light penetrates into the interior ($x > 0$) of the bar. Determine $\Delta p_n(x)$.

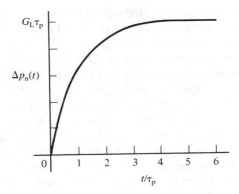

Fig. 3.17 Solution to Sample Problem No. 1. Photogeneration-induced increase in the excess hole concentration as a function of time.

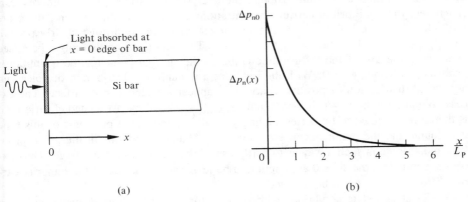

(a)   (b)

Fig. 3.18 (a) Pictorial definition of Sample Problem No. 2. (b) Solution to Sample Problem No. 2 showing the excess hole concentration inside the Si bar as a function of position.

Solution: The semiconductor is again silicon uniformly doped with an $N_D = 10^{15}/cm^3$. Steady state conditions are inferred from the statement of the problem, since we are asked for $\Delta p_n(x)$ and not $\Delta p_n(x, t)$. Moreover, at $x = 0$, $\Delta p_n(0) = \Delta p_{n0} = 10^{10}/cm^3$, and $\Delta p_n \to 0$ as $x \to \infty$. The latter boundary condition follows from the semi-infinite nature of the bar. The perturbation in Δp_n due to the nonpenetrating light can't possibly extend out to $x = \infty$. The nonpenetrating nature of the light itself tells us that $G_L = 0$ for $x > 0$. Finally, note that the problem statement fails to mention the temperature of operation. When this happens it is reasonable to assume an intended $T = $ room temperature.

If we take $T =$ room temperature, the silicon bar in Problem 2 would revert to an equilibrium condition identical to that described in Problem 1 if the light were removed. Under equilibrium conditions, then, $n_0 = 10^{15}/cm^3$, $p_0 = 10^5/cm^3$, and the carrier concentrations are uniform throughout the semiconductor bar.

Qualitatively it is a simple matter to predict the expected effect of the nonpenetrating light on the silicon bar. The light first creates excess carriers right at $x = 0$. With more carriers at one point than elsewhere in the bar, the diffusion process next comes into play and the carrier excess spreads into the semiconductor proper. At the same time, however, the appearance of an excess hole concentration inside the bar enhances the thermal recombination rate. Thus, as the diffusing holes move into the bar their numbers are reduced by recombination. In addition, since the minority carrier holes live for only a limited time period, a time τ_p on the average, fewer and fewer excess holes will survive as x becomes larger and larger. Under steady state conditions it is therefore reasonable to expect an excess distribution of holes near $x = 0$, with $\Delta p_n(x)$ monotonically decreasing from Δp_{n0} at $x = 0$ to $\Delta p_n = 0$ as $x \to \infty$.

In preparation for obtaining a quantitative solution we observe that the system under consideration is one-dimensional, the analysis is restricted to the minority carrier holes, the equilibrium carrier concentrations are position-independent, there are no "other processes" for $x > 0$, and low level injection conditions clearly prevail ($\Delta p_{n|max} = \Delta p_{n0} = 10^{10}/cm^3 \ll n_0 = 10^{15}/cm^3$). The only question that might be raised concerning the use of the diffusion equation as the starting point for the quantitative analysis is whether $\mathscr{E} \simeq 0$. With the light on, a nonuniform distribution of holes and associated distribution of positive charge will appear near the $x = 0$ surface. The excess hole pile-up, however, is very small ($\Delta p_{n|max} \simeq n_i$) and the associated electric field is therefore expected to be correspondingly small. Moreover, in problems of this type it is found that the majority carriers, negatively charged electrons in the given problem, redistribute in such a way as to partly cancel the minority carrier charge. Thus experience indicates the $\mathscr{E} \simeq 0$ assumption to be reasonable, and use of the minority carrier diffusion equation to be justified.

Under steady state conditions with $G_L = 0$ for $x > 0$ the hole diffusion equation reduces to the form

$$D_P \frac{d^2 \Delta p_n}{dx^2} - \frac{\Delta p_n}{\tau_p} = 0 \qquad \text{for } x > 0 \tag{3.51}$$

which is to be solved subject to the boundary conditions

$$\Delta p_{n|x=0^+} = \Delta p_{n|x=0} = \Delta p_{n0} \tag{3.52}$$

and

$$\Delta p_{n|x \to \infty} = 0 \tag{3.53}$$

Equation (3.51) should be recognized as one of the simplified diffusion equations cited in Subsection 3.4.3, with the general solution

$$\Delta p_n(x) = Ae^{-x/L_P} + Be^{x/L_P} \tag{3.54}$$

where

$$L_P \equiv \sqrt{D_P \tau_P} \tag{3.55}$$

Because $\exp(x/L_P) \to \infty$ as $x \to \infty$, the only way that the Eq. (3.53) boundary condition can be satisfied is for B to be identically zero. With $B = 0$, application of the Eq. (3.52) boundary condition yields

$$A = \Delta p_{n0} \tag{3.56}$$

and

$$\Delta p_n(x) = \Delta p_{n0} e^{-x/L_P} \quad \Leftarrow \text{solution} \tag{3.57}$$

The Eq. (3.57) result is plotted in Fig. 3.18(b). In agreement with qualitative arguments, the nonpenetrating light merely gives rise to a monotonically decreasing $\Delta p_n(x)$ starting from Δp_{n0} at $x = 0$ and decreasing to $\Delta p_n = 0$ as $x \to \infty$. Note that the precise functional form of the falloff in the excess carrier concentration is exponential with a characteristic decay length equal to L_P.

3.5.3 Diffusion Lengths

The situation just encountered in Sample Problem No. 2 — the creation (or appearance) of an excess of minority carriers along a given plane in a semiconductor, the subsequent diffusion of the excess from the point of injection, and the exponential falloff in the excess carrier concentration characterized by a decay length (L_P) — occurs often enough in semiconductor analyses that the characteristic length has been given a special name. Specifically,

$$L_P \equiv \sqrt{D_P \tau_P} \qquad \begin{array}{l}\text{associated with the minority carrier} \\ \text{holes in an } n\text{-type material}\end{array} \tag{3.58a}$$

and

$$L_N \equiv \sqrt{D_N \tau_n} \qquad \begin{array}{l}\text{associated with the minority carrier} \\ \text{electrons in a } p\text{-type material}\end{array} \tag{3.58b}$$

are referred to as *minority carrier diffusion lengths*.

Physically, L_P and L_N represent the average distance minority carriers can diffuse into a sea of majority carriers before being annihilated. This interpretation is clearly consistent with Sample Problem No. 2, where the average position of the excess minority carriers inside the semiconductor bar is

$$\langle x \rangle = \int_0^\infty x \Delta p_n(x)\, dx \Big/ \int_0^\infty \Delta p_n(x)\, dx = L_P \tag{3.59}$$

For memory purposes, minority carrier diffusion into a sea of majority carriers might be likened to a small group of animals attempting to cross a piranha-infested stretch of the Amazon River. In the analogy L_P and L_N correspond to the average distance the animals advance into the river before being eaten.

To obtain some idea as to the size of diffusion lengths, let us assume $T =$ room temperature, an $N_D = 10^{15}/\text{cm}^3$ doped semiconductor, and a $\tau_p = 10^{-6}$ sec. For the given system

$$L_P = \sqrt{D_P \tau_p} = \sqrt{(kT/q)\mu_p \tau_p} = [(0.026)(458)(10^{-6})]^{1/2}$$
$$\simeq 3.5 \times 10^{-3} \text{ cm}$$

3.5.4 Quasi-Fermi Levels

Quasi-Fermi levels are energy levels used to specify the carrier concentrations inside a semiconductor under *non*equilibrium conditions.

To understand the need for introducing quasi-Fermi levels let us first refer to Sample Problem No. 1. In this problem equilibrium conditions prevailed prior to $t = 0$, with $n_0 = N_D = 10^{15}/\text{cm}^3$ and $p_0 = 10^5/\text{cm}^3$. The energy band diagram describing the equilibrium situation is shown in Fig. 3.19(a). A simple inspection of the energy band diagram and the Fermi level positioning also conveys, of course, the equilibrium carrier concentrations, since

$$n_0 = n_i e^{(E_F - E_i)/kT} \tag{3.60a}$$

$$p_0 = n_i e^{(E_i - E_F)/kT} \tag{3.60b}$$

The point we wish to emphasize is that under equilibrium conditions there is a one-to-one correspondence between the Fermi level and the carrier concentrations. Knowledge of E_F completely specifies n_0 and p_0 and vice versa.

Let us next turn to the nonequilibrium (steady state) situation inside the Problem 1 semiconductor at times $t \gg \tau_p$. For times $t \gg \tau_p$, $\Delta p_n = G_L \tau_p = 10^{11}/\text{cm}^3$, $p = p_0 + \Delta p \simeq 10^{11}/\text{cm}^3$ and $n \simeq n_0 = 10^{15}/\text{cm}^3$. Although n remains essentially unperturbed, p has increased by many orders of magnitude; and it is clear that the Fig. 3.19(a) diagram no longer describes the state of the system. In fact, the Fermi level is defined

Fig. 3.19 Sample use of quasi-Fermi levels. Energy band description of the situation inside the semiconductor of Sample Problem No. 1 under (a) equilibrium conditions and (b) nonequilibrium conditions ($t \gg \tau_p$).

only for a system under equilibrium conditions and cannot be used to deduce the carrier concentrations inside a system in a nonequilibrium state.

The convenience of being able to deduce the carrier concentrations by inspection from the energy band diagram is extended to nonequilibrium conditions through the use of *quasi-Fermi levels*. This is accomplished by introducing two energies, F_N, the quasi-Fermi level for electrons, and F_P, the quasi-Fermi level for holes, which are *by definition* related to the nonequilibrium carrier concentrations in the same way E_F is related to the equilibrium carrier concentrations. To be specific, under nonequilibrium conditions and assuming the semiconductor to be nondegenerate,

$$n \equiv n_i e^{(F_N - E_i)/kT} \quad \text{or} \quad F_N \equiv E_i + kT \ln\left(\frac{n}{n_i}\right) \tag{3.61a}$$

and

$$p \equiv n_i e^{(E_i - F_P)/kT} \quad \text{or} \quad F_P \equiv E_i - kT \ln\left(\frac{p}{n_i}\right) \tag{3.61b}$$

Please note that F_N and F_P are conceptual constructs, with the values of F_N and F_P being totally determined from a prior knowledge of n and p. Moreover, the quasi-Fermi level definitions have been carefully chosen so that when a perturbed system relaxes back toward equilibrium, $F_N \to E_F$, $F_P \to E_F$ and Eqs. (3.61) → Eqs. (3.60).

To provide a straightforward application of the quasi-Fermi level formalism, let us again consider the $t \gg \tau_p$ state of the semiconductor in Sample Problem No. 1. First, since $n \simeq n_0$, $F_N \simeq E_F$. Next, substitution of $p = 10^{11}/\text{cm}^3$ ($n_i = 10^{10}/\text{cm}^3$ and $kT = 0.026$ eV) into Eq. (3.61b) yields $F_P = E_i - 0.06$ eV. By eliminating E_F from the energy band diagram and drawing lines at the appropriate energies to represent F_N and F_P, one obtains the diagram displayed in Fig. 3.19(b). Figure 3.19(b) clearly conveys to any observer that the system under analysis is in a nonequilibrium state. Figure 3.19(b) referenced to Fig. 3.19(a) further indicates at a glance that low level injection is taking place inside the semiconductor creating a concentration of minority carrier holes in excess of n_i. A second application based on Sample Problem No. 2, an application where the quasi-Fermi levels are a function of position, is left to the reader as an exercise.

As a final point, it should be mentioned that the quasi-Fermi level formalism can be used to recast some of the carrier-action relationships in a more compact form. For example, the standard form of the equation for the total hole current reads

$$\mathbf{J}_P = q\mu_p p\mathscr{E} - qD_P\nabla p \tag{3.62}$$

(same as 3.42b)

Differentiating both sides of Eq. (3.61b) with respect to position, one obtains

$$\nabla p = \left(\frac{n_i}{kT}\right)e^{(E_i - F_P)/kT}(\nabla E_i - \nabla F_P) \tag{3.63a}$$

$$= \left(\frac{qp}{kT}\right)\mathscr{E} - \left(\frac{p}{kT}\right)\nabla F_P \tag{3.63b}$$

The identity $\mathscr{E} = \nabla E_i/q$ [the three-dimensional version of Eq. (3.15)] is employed in progressing from Eq. (3.63a) to Eq. (3.63b). Next eliminating ∇p in Eq. (3.62) using Eq. (3.63b) gives

$$\mathbf{J}_P = q\left(\mu_p - \frac{qD_P}{kT}\right)p\mathscr{E} + \left(\frac{qD_P}{kT}\right)p\nabla F_P \tag{3.64}$$

From the Einstein relationship, however, $qD_P/kT = \mu_p$. We therefore conclude

$$\mathbf{J}_P = \mu_p p\nabla F_P \tag{3.65a}$$

Similarly,

$$\mathbf{J}_N = \mu_n n\nabla F_N \tag{3.65b}$$

Since $\mathbf{J}_P \propto \nabla F_P$ and $\mathbf{J}_N \propto \nabla F_N$ in Eqs. (3.65), one is led to a very interesting general interpretation of energy band diagrams containing quasi-Fermi levels. Namely, a quasi-Fermi level that varies with position ($dF_P/dx \neq 0$ or $dF_N/dx \neq 0$) indicates at a glance that current is flowing inside the semiconductor.

SEE EXERCISE 3.2– APPENDIX A

3.6 REVIEW AND CONCLUDING COMMENTS

The examination of sample problems concludes the discussion on carrier action in particular and semiconductor fundamentals in general. Over the course of the volume, the material covered has dealt with a wide variety of items including terms, relevant facts and information, representative numerical values, models or visualization aids, and working equations. With little hesitation the reader should be able to define such terms as unit cell, donor, p-type material, degeneracy, scattering, R–G center, and diffusion length. A more complete collection of fundamental terms is listed in Table 3.1. Useful facts and information were scattered throughout the development. For example, the crystalline structure of Si is described by the diamond lattice, insulators have very large band gaps, and in the major semiconductors $\mu_n > \mu_p$. More extensive information was presented concerning the Miller indexing scheme, the determination of doping type using the hot-point probe, and the measurement of the doping concentration using the four-point probe. A different kind of knowledge, knowledge of representative numerical values, is invaluable in performing crude order-of-magnitude calculations and in gauging the relative size of newly encountered quantities. Table 3.2 lists numerical values for a selected set of key parameters.

The prime visualization aid described herein was, of course, the energy band diagram. The full description of the diagram actually spanned most of the volume. The basic diagram introduced in Section 2.2 consisted of only two energy levels, E_c and E_v. Donor and acceptor levels were then added to the diagram and were used to help explain the action of dopants in modifying the carrier concentrations. Next the energy levels E_F and E_i were placed on the diagram to indicate at a glance the equilibrium carrier concentrations. Later it was pointed out that the existence of an electric field inside the semiconductor caused band bending or a variation of the energy bands with position. By simply inspecting the diagram one could ascertain the general functional dependence of the electrostatic potential and electric field present in the material. In the discussion of recombination–generation another level was added near midgap. This level arising from R–G centers plays a dominant role in the thermal communication between the bands. Lastly, quasi-Fermi levels were introduced to describe nonequilibrium conditions.

Working equations developed or derived herein fall into two general categories. One set of equations is used to quantitatively characterize the equilibrium state of the semiconductor. The most important of these equations are cited in the summary at the end of Chapter 2. The second group of equations is used to describe carrier action or model the state of a semiconductor subject to an external perturbation. The latter "equations of state" are reviewed and listed in Subsection 3.4.4. In dealing with either equation set one must always be careful to check the applicability of the equations prior to their utilization.

In conclusion, we have presented and examined those terms, facts, models, equations, etc., that are routinely encountered in semiconductor-related analyses; we have

Table 3.1   List of Terms.

(1) amorphous	(31) extrinsic temperature region
(2) polycrystalline	(32) intrinsic temperature region
(3) crystalline	(33) freeze-out
(4) lattice	(34) drift
(5) unit cell	(35) scattering
(6) ingot	(36) drift velocity
(7) carrier	(37) thermal motion
(8) electron	(38) drift current
(9) hole	(39) current density
(10) conduction band	(40) mobility
(11) valence band	(41) resistivity
(12) band gap	(42) conductivity
(13) effective mass	(43) band bending
(14) intrinsic semiconductor	(44) diffusion
(15) extrinsic semiconductor	(45) diffusion current
(16) dopant	(46) diffusion coefficient
(17) donor	(47) recombination
(18) acceptor	(48) generation
(19) n-type material	(49) photogeneration
(20) p-type material	(50) direct thermal R–G
(21) n^+ (or p^+) material	(51) R–G centers
(22) majority carrier	(52) R–G via R–G centers
(23) minority carrier	(53) low level injection
(24) density of states	(54) equilibrium
(25) Fermi function	(55) perturbation
(26) Fermi energy (or level)	(56) steady state
(27) nondegenerate semiconductor	(57) quasi steady state
(28) degenerate semiconductor	(58) minority carrier lifetime
(29) charge neutrality	(59) minority carrier diffusion length
(30) ionization of dopant sites	(60) quasi-Fermi level

Table 3.2 Numerical Values for Key Parameters.

Quantity	Approximate Value (Si, $T = 300$ K)
kT	0.026 eV
E_G	1.12 eV
n_i	$10^{10}/cm^3$
μ_n and μ_p	1360 and 460 cm^2/V-sec (for N_D or $N_A = 10^{14}/cm^3$)
τ_n and τ_p	$\sim 10^{-6}$ sec (actual values can range from 10^{-3} to 10^{-9} sec)
L_N and L_P	$\sim 10^{-2}$–10^{-3} cm

sought to lay a firm foundation for the reader's future adventures into the world of solid state devices.

PROBLEMS

3.1 Compile an "Abbreviated Dictionary of Semiconductor Terminology" based on the list of terms collected in Table 3.1. Preferably use your own words in defining the various terms.

3.2 Using the energy band diagram, indicate how one visualizes:

(a) The existence of an electric field inside a semiconductor;

(b) An electron with a K. E. $= 0$;

(c) A hole with a K. E. $= E_G/4$;

(d) Photogeneration;

(e) Direct thermal generation;

(f) Direct thermal recombination;

(g) Recombination via R–G centers;

(h) Generation via R–G centers.

3.3 Short Answers

(a) An average hole drift velocity of 10^3 cm/sec results when 2 V is applied across a 1-cm-long semiconductor bar. What is the hole mobility inside the bar?

(b) Name the two dominant carrier *scattering* mechanisms in nondegenerately doped semiconductors of device quality.

(c) For a given semiconductor the carrier mobilities in intrinsic material are (choose one: higher than, lower than, the same as) those in heavily doped material. Briefly explain why the mobilities in intrinsic material are (chosen answer) those in heavily doped material.

(d) Two bars of the same material, one n-type and one p-type, are uniformly doped such that N_D(bar 1) = N_A(bar 2) $\gg n_i$. Which bar will exhibit the larger resistivity? Explain.

(e) The electron mobility in a silicon sample is determined to be 1300 cm^2/V-sec at room temperature. What is the electron diffusion coefficient?

(f) What is the algebraic statement of low level injection?

(g) Light is used to create excess carriers in silicon. These excess carriers will predominantly recombine via (choose one: direct thermal, R–G center, or photo) recombination.

(h) Prior to processing, a semiconductor contains $N_D = 10^{14}$/cm^3 donors and $N_T = 10^{11}$/cm^3 R–G centers. After processing (say in the fabrication of a device), the semiconductor sample contains $N_D = 10^{17}$/cm^3 donors and $N_T = 10^{10}$/cm^3 R–G centers. Did processing increase or decrease the minority carrier lifetime? Explain.

3.4 Resistivity

(a) A silicon sample maintained at room temperature is uniformly doped with $N_D = 10^{16}$/cm^3 donors. Calculate the resistivity of the sample using Eq. (3.8a). Compare your calculated result with the ρ deduced from Fig. 3.7.

(b) The silicon sample of part (a) is "compensated" by adding $N_A = 10^{16}$/cm^3 acceptors. Calculate the resistivity of the compensated sample. (Exercise caution in choosing the mobility values to be employed in this part of the problem.)

(c) Compute the resistivity of intrinsic ($N_A = 0$, $N_D = 0$) silicon at room temperature. How does your result here compare with that for part (b)?

(d) A 500-ohm resistor is to be made from a bar-shaped piece of n-type Si. The bar has a cross-sectional area of 10^{-2} cm^2 and a current-carrying length of 1 cm. Determine the doping required.

3.5 Fill in the following table by deducing the required information from the energy band diagram of Fig. P3.5.

Carrier	K. E. (eV)	P. E. (eV)
Electron 1		
Electron 2		
Electron 3		
Hole 1		
Hole 2		
Hole 3		

Fig. P3.5

3.6 (a) Construct a rough plot of D_N versus doping appropriate for $10^{14}/\text{cm}^3 \leq N_D$ or $N_A \leq 10^{19}/\text{cm}^3$ doped Si maintained at room temperature. Record the reasoning underlying your plot construction.

(b) Construct a rough plot of D_P versus T appropriate for *lightly doped* p-type silicon. Restrict your plot to temperatures near room temperature and record the reasoning leading to your plot.

3.7 A semiconductor under equilibrium conditions is characterized by the energy band diagram of Fig. P3.7. Given $E_G = 1.12$ eV, $n_i = 10^{10}/\text{cm}^3$, and $kT = 0.0259$ eV,

(a) Determine n at $x = L/4$, $x = L/2$, and $x = 3L/4$.

(b) For what values of x, if any, is the semiconductor degenerate?

(c) Sketch n and p versus x.

(d) Sketch V inside the semiconductor as a function of x.

(e) Sketch $\mathscr{E}$ inside the semiconductor as a function of x.

(f) If $L = 10^{-2}$ cm, what is the magnitude of $\mathscr{E}$ at $x = L/2$?

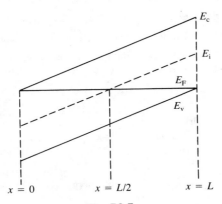

Fig. P3.7

3.8 A semiconductor device is characterized by the idealized energy band diagram in Fig. P3.8. It is also known that $E_G = 1.12$ eV, $n_i = 10^{10}/cm^3$, and $kT = 0.0259$ eV; $\mu_n = 1345$ cm^2/V-sec and $\mu_p = 458$ cm^2/V-sec at both $x = x_a$ and $x = x_b$.

(a) Sketch the electrostatic potential (V) inside the semiconductor as a function of x. Set $V = 0$ at $x = x_a$.

(b) Sketch the electric field ($\mathscr{E}$) inside the semiconductor as a function of x.

(c) What is the magnitude of the electron drift current density ($J_{N|drift}$) at $x = x_a$? Explain.

(d) Is there an electron drift current at $x = x_b$? If there is a drift current, what is the direction of the current flow?

(e) Is there an electron diffusion current at $x = x_b$? If there is a diffusion current, what is the direction of the current flow?

(f) What is the magnitude of the total current density (J) at $x = x_b$? Explain.

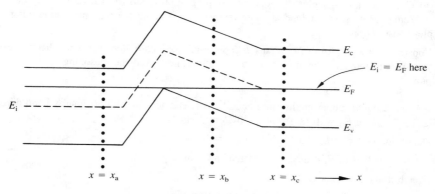

Fig. P3.8

3.9 A semiconductor is characterized by the energy band diagram in Fig. P3.9. It is also known that $E_G = 1.12$ eV, $kT = 0.0259$ eV, $n_i = 10^{10}/cm^3$, $\mu_n = 1350$ cm^2/V-sec, and $\tau_n = 10^{-4}$ sec.

(a) Sketch the electrostatic potential (V) inside the semiconductor as a function of x.

(b) Sketch the electric field ($\mathscr{E}$) inside the semiconductor as a function of x.

(c) For what values of x does charge neutrality exist inside the semiconductor? (HINT: from Poisson's equation $d\mathscr{E}/dx = \rho/K_S\varepsilon_0$, where $\rho = $ charge/cm^3.)

(d) Sketch the electron concentration as a function of position. Specifically indicate on your plot the values of n at $x = x_a$ and $x = x_c$.

(e) (i) Is there an electron drift current at $x = x_a$?
   (ii) Is there an electron diffusion current at $x = x_a$?
   (iii) What is the total electron current density (J_N) at $x = x_a$? Explain your answers.

(f) An electron at $x = x_b$ with a total energy $E \simeq E_c$ moves from $x = x_b$ to $x = 0$ without changing its total energy. What is the kinetic energy (K. E.) of the electron upon arriving at $x = 0$?

(g) A small excess of electrons is introduced at the point x_c. The excess electrons diffuse into the semiconductor toward $x = x_b$. If $x_c - x_b = 10^{-3}$ cm, will a significant number (say $\geq 10\%$) of the excess electrons ever reach x_b? Explain.

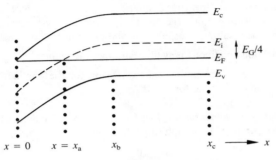

Fig. P3.9

3.10 Show that, under steady state conditions and low level injection,

$$\Delta p_n(x) = \Delta p_{n0}(1 - x/L) \qquad \ldots 0 \leq x \leq L$$

is the special-case solution of the minority carrier diffusion equation which will result if (1) one assumes ALL recombination–generation processes are negligible within an n-type semiconductor of length L, and (2) one employs the boundary conditions $\Delta p_n(0) = \Delta p_{n0}$ and $\Delta p_n(L) = 0$. (Neglecting recombination–generation is an excellent approximation when L is much less than a minority carrier diffusion length. Solutions of the type presented here are frequently encountered in practical problems.)

3.11 A silicon wafer ($N_A = 10^{14}/\text{cm}^3$, $\tau_n = 1\ \mu\text{sec}$, T = room temperature) is first illuminated for a time $t \gg \tau_n$ with light which generates $G_{L0} = 10^{16}$ electron-hole pairs per cm^3-sec uniformly throughout the volume of the silicon. At time $t = 0$ the light intensity is reduced, making $G_L = G_{L0}/2$ for $t \geq 0$. Determine $\Delta n_p(t)$ for $t \geq 0$.

3.12 A semi-infinite p-type bar (see Fig. P3.12) is illuminated with light which generates G_L electron-hole pairs per cm^3-sec uniformly throughout the volume of the semiconductor. Simulta-

Fig. P3.12

neously, carriers are extracted at $x = 0$ making $\Delta n_p = 0$ at $x = 0$. Assuming a steady state condition has been established and $\Delta n_p(x) \ll p_0$ for all x, determine $\Delta n_p(x)$.

3.13 In a uniformly doped Si sample maintained at room temperature, $N_D = 10^{14}/\text{cm}^3$ and $\tau_p = 10^{-6}$ sec. The sample is first illuminated for a time $t \gg \tau_p$ with light that generates $G_L = 10^{16}$ electron-hole pairs per cm^3-sec uniformly throughout the silicon. At time $t = 0$ the light is turned off and the resulting transient observed as a function of time.

(a) Do low level injection conditions prevail at all times? Explain.

(b) Assuming $\Delta n(t) = \Delta p(t)$, establish an expression for the conductivity of the sample ($\sigma = 1/\rho$) as a function of time for $t \geq 0$.

3.14 A Si bar of length L is *nonuniformly* doped with donors and maintained under *equilibrium* conditions at room temperature. It is established that

$$n(x) = n_i e^{(x+a)/b} \qquad \ldots 0 \leq x \leq L$$

where $a = 0.5$ cm, $b = 0.1$ cm, and $L = 1$ cm.

(a) Draw the energy band diagram for the bar specifically showing E_c, E_F, E_i, and E_v on your diagram. Explain how you arrived at your diagram.

(b) Make a sketch of the electric field ($\mathscr{E}$) inside the bar as a function of position.

(c) What are the respective directions ($+x$ or $-x$) of the electron drift, electron diffusion, hole drift, and hole diffusion currents flowing inside the bar?

3.15 An *n*-type Si bar of length L (where $0 \leq x \leq L$) is maintained under steady state conditions such that $\Delta p_n(0) = \Delta p_{n0} = 10^{12}/\text{cm}^3$ and $\Delta p_n(L) = 0$. The Si bar is uniformly doped with $N_D = 10^{16}/\text{cm}^3$ donors, $T =$ room temperature, and there is no photogeneration or "other processes" of any type occurring inside the bar.

(a) Do low level injection conditions prevail? Explain.

(b) $n(x) = ?$

(c) What differential equation (simplest form possible) must be solved to determine $\Delta p_n(x)$ inside the bar?

(d) What is the position of the hole quasi-Fermi level inside the bar at (i) $x = 0$ and (ii) $x = L$?

3.16 The equilibrium and steady state conditions before and after illumination of a semiconductor are characterized by the energy band diagrams shown in Fig. P3.16. $T = 300$ K, $n_i = 10^{10}/\text{cm}^3$, $\mu_n = 1345$ cm^2/V-sec, and $\mu_p = 458$ cm^2/V-sec. From the information provided, determine:

(a) n_0 and p_0, the equilibrium carrier concentrations;

(b) n and p under steady state conditions;

(c) N_D.

(d) Do we have "low level injection" when the semiconductor is illuminated? Explain.

(e) What is the resistivity of the semiconductor before and after illumination?

(a) Before   (b) After

Fig. P3.16

3.17 A portion ($0 \leq x \leq L$) of a Si sample, uniformly doped with $N_D = 10^{15}/\text{cm}^3$ donors and maintained at room temperature, is subject to a *steady state* perturbation such that

$$n \simeq N_D$$
$$p = n_i(1 - x/L) + n_i^2/N_D \qquad \ldots 0 \leq x \leq L$$

Since $n \simeq N_D$, it is reasonable to assume $\mathscr{E} \simeq 0$ in the $0 \leq x \leq L$ region. Given $\mathscr{E} \simeq 0$, sketch the energy band diagram for the perturbed region specifically including E_c, E_i, E_v, F_N, and F_P on your diagram.

Suggested Readings

General comment: Most texts devoted to a general coverage of solid state devices contain an introductory set of chapters on semiconductor fundamentals. Listed below are six texts which, in the author's opinion, best supplement or enhance the material presented in this volume. These texts contain additional references that might also be consulted by the reader.

R. B. Adler, A. C. Smith, and R. L. Longini, *Introduction to Semiconductor Physics,* Semiconductor Electronics Education Committee (SEEC) Vol 1, New York: Wiley, 1964. SEEC Volume 1 is a work very similar in scope and content to the present volume. It is a bit difficult to read and slightly more advanced in spots, but is generally considered a classic in its field.

D. H. Navon, *Semiconductor Microdevices and Materials*. New York: Holt, Rinehart and Winston, 1986. See Chapters 2 to 5. The Navon text contains a more advanced (but still understandable) coverage of the material presented in this volume. It is a text for readers who wish to expand their knowledge of the subject matter.

R. F. Pierret, *Advanced Semiconductor Fundamentals,* Vol VI in the Modular Series on Solid State Devices, R. F. Pierret and G. W. Neudeck, editors. Reading, MA: Addison-Wesley, 1987. Volume VI is primarily intended for use in advanced undergraduate and beginning graduate level studies. It is an excellent text for those wishing a greater depth of understanding and a broader coverage of semiconductor fundamentals.

B. G. Streetman, *Solid State Electronic Devices*. 2nd Ed. Englewood Cliffs, NJ: Prentice-Hall, 1980. See Chapters 1 to 4. The book by Streetman is a good general-purpose undergraduate text providing a different point of view on many of the topics covered in Volume I.

S. M. Sze, *Semiconductor Devices, Physics and Technology*. New York: Wiley, 1985. See Chapters 1 and 2. The Sze text provides a concise coverage of semiconductor fundamentals at a level intermediate between Volumes I and VI of the Modular Series.

H. E. Talley and D. G. Daugherty, *Physical Principles of Semiconductor Devices*. Ames: Iowa State University Press, 1976. See Chapters 1 to 6. This is a fairly readable text that provides a quantum-mechanical and statistical background for some of the concepts and relationships simply introduced herein.

Appendix A
Exercises

EXERCISE 1.1

Q: If the lattice constant or unit cell side-length in Si is $a = 5.43 \times 10^{-8}$ cm, what is the distance (d) from the center of one Si atom to the center of its nearest neighbor?

A: As noted in the Fig. 1.4 caption, the atom in the upper front corner of the Si unit cell and the atom along the cube diagonal one-fourth of the way down the diagonal are nearest neighbors. Since the diagonal of a cube is equal to $\sqrt{3}$ times a side length, one concludes $d = (1/4)\sqrt{3}a = (\sqrt{3}/4)(5.43 \times 10^{-8}) = \boxed{2.35 \times 10^{-8} \text{ cm}}$.

EXERCISE 1.2

Q: For a cubic crystal lattice:

(a) Determine the Miller indices for the plane and direction vector shown in Fig. E1.2(a).

(b) Sketch the plane and direction vector characterized by (110) and [110], respectively.

A: (a) The intercepts of the plane are -1, 1, and 2 along the x, y, and z coordinate axes, respectively. Taking [1/intercept]s yields $-1, 1, \frac{1}{2}$. Multiplying by 2 to obtain the lowest whole number set, and enclosing in parentheses, gives $\boxed{(\bar{2}21)}$ for the Miller indices of the plane. The direction vector has projections of $2a$, a, and 0 along the x, y, and z coordinate axes, respectively. The Miller indices for the direction are then $\boxed{[210]}$.

(b) For the (110) plane, [1/intercept]s $= 1, 1, 0$. The normalized x, y, z axes intercepts of the plane are therefore $1, 1, \infty$; the plane intersects the x-axis at a, the y-axis at a, and is parallel to the z-axis. In a cubic crystal the [110] direction is normal to the (110) plane. The deduced plane and direction are as sketched in Fig. E1.2(b).

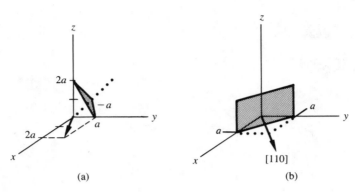

Fig. E1.2

EXERCISE 2.1

Energy Quiz

Q: (a) 1 eV is equal to how many joules of energy?

(b) kT is equal to how many eV at 300 K?

(c) The ionization energy of acceptors and donors in Si is roughly equal to __?__

(d) $E_G(Si)$ = __?__

(e) $E_G(SiO_2)$ = __?__

(f) The energy required to ionize a hydrogen atom initially in the **n** = 1 state is __?__

A: (a) 1 eV = 1.6×10^{-19} joules.

(b) kT = 0.0259 eV ($\simeq$ 0.026 eV) at 300 K.

(c) The ionization energy of dopants is equal to $|E_B|$ of the dopant sites. As discussed in Subsection 2.3.4, $|E_B| \simeq 0.1$ eV for dopants in Si.

(d) $E_G(Si)$ = 1.12 eV at room temperature (see Subsection 2.2.4).

(e) $E_G(SiO_2) \simeq 8$ eV (see Subsection 2.2.4).

(f) Making use of Eq. (2.1) in Section 2.1, the hydrogen atom ionization energy is $|E_{H|n=1}|$ = 13.6 eV.

EXERCISE 2.2

Q: The probability that a state is filled at the conduction band edge (E_c) is precisely equal to the probability that a state is *empty* at the valence band edge (E_v). Where is the Fermi level located?

A: The Fermi function, $f(E)$, specifies the probability of electrons occupying states at a given energy E. The probability that a state is empty (not filled) at a given energy E is equal to $1 - f(E)$. Here we are told

$$f(E_c) = 1 - f(E_v)$$

Since

$$f(E_c) = \frac{1}{1 + e^{(E_c - E_F)/kT}}$$

and

$$1 - f(E_v) = 1 - \frac{1}{1 + e^{(E_v - E_F)/kT}} = \frac{1}{1 + e^{(E_F - E_v)/kT}}$$

we conclude

$$(E_c - E_F)/kT = (E_F - E_v)/kT$$

or

$$\boxed{E_F = (E_c + E_v)/2}$$

The Fermi level is positioned at midgap.

EXERCISE 2.3

Q: A Si sample is doped with 10^{14} boron atoms per cm^3.

(a) What are the carrier concentrations in the Si sample at 300 K?

(b) What are the carrier concentrations at 470 K?

A: (a) Boron in Si is an acceptor (see Table 2.2). Thus $N_A = 10^{14}/cm^3$. At 300 K, $n_i = 1.18 \times 10^{10}/cm^3$ and the given N_A is clearly much greater than n_i. Moreover, since the N_D doping was omitted from the problem statement, we infer $N_D \ll N_A$. With $N_A \gg n_i$ and $N_A \gg N_D$, Eqs. (2.31) may be used to calculate the carrier concentrations: $p = N_A = 10^{14}/cm^3$; $n = n_i^2/N_A = 1.39 \times 10^6/cm^3$.

(b) As deduced from Fig. 2.20, $n_i = 10^{14}/cm^3$ at 470 K. Because n_i is comparable to N_A, Eqs. (2.29) must be used to calculate at least one of the carrier concentrations. (Once one of the carrier concentrations is known, the second is more readily computed using the np product expression.) Performing the indicated calculations gives: $p = N_A/2 + [(N_A/2)^2 + n_i^2]^{1/2} = 1.62 \times 10^{14}/\text{cm}^3$; $n = n_i^2/p = 6.18 \times 10^{13}/\text{cm}^3$.

EXERCISE 2.4

Q: For each of the conditions specified in Exercise 2.3, determine the position of E_i, compute $E_F - E_i$, and draw a carefully dimensioned energy band diagram for the Si sample. Note that $E_G(\text{Si}) = 1.08$ eV at 470 K and assume m_p^*/m_n^* is temperature independent.

A: (a) In part (a) of Exercise 2.3 the $N_A = 10^{14}/\text{cm}^3$ doped Si sample is maintained at 300 K. Using Eq. (2.36) we conclude E_i is located 0.0073 eV below midgap. [Actually, the positioning of E_i in Si at 300 K is noted in the text below Eq. (2.36).] Next applying Eq. (2.38b), we find

$$E_i - E_F \simeq kT \ln(N_A/n_i)$$
$$= 0.0259 \ln(10^{14}/1.18 \times 10^{10})$$
$$= 0.234 \text{ eV}$$

The energy band diagram constructed from the deduced positioning of E_i and E_F is shown in Fig. E2.4(a).

(b) In part (b) of Exercise 2.3 the Si sample is heated to 470 K. With m_p^*/m_n^* assumed to be temperature independent and $kT = 0.0405$ eV at 470 K, $(3/4)kT \ln(m_p^*/m_n^*) = -0.0114$ eV and E_i is deduced to be located 0.0114 eV below midgap. Because N_A is comparable to n_i at 470 K, Eq. (2.37) must be used to compute the positioning of E_F. Specifically, with $n_i = 10^{14}/\text{cm}^3$ and $p = 1.62 \times 10^{14}/\text{cm}^3$ (see Exercise 2.3),

$$E_i - E_F = kT \ln(p/n_i)$$
$$= 0.0405 \ln(1.62 \times 10^{14}/10^{14})$$
$$= 0.0195 \text{ eV}$$

Here E_F is only slightly removed from E_i as pictured in Fig. E2.4(b).

(a)   (b)

Fig. E2.4

EXERCISE 3.1

Q: A semiconductor is characterized by the energy band diagram of Fig. E3.1.

Fig. E3.1

(a) If the semiconductor is taken to be Si maintained at $T = 300$ K ($E_G = 1.12$ eV, $n_i = 1.18 \times 10^{10}/\text{cm}^3$, and $kT = 0.0259$ eV), determine the resistivity of the $x > W/2$ portion of the semiconductor.

(b) An electron located at $x = W/2$ attempts to move over to the $x < -W/2$ region of the semiconductor without changing its total energy. What is the minimum kinetic energy the electron must have in order to accomplish this feat?

(c) Sketch the electrostatic potential (V) inside the semiconductor as a function of x.

(d) Sketch the electric field ($\mathscr{E}$) inside the semiconductor as a function of x.

(e) The semiconductor is in equilibrium. How does one deduce this fact from the energy band diagram?

(f) What is the electron current density (J_N) and hole current density (J_P) at $x = 0$?

(g) Is there an electron *drift* current at $x = 0$? If there is a drift current, what is the direction of the current flow?

(h) Is there an electron *diffusion* current at $x = 0$? If there is a diffusion current, what is the direction of the current flow?

A: (a) The $x > W/2$ portion of the semiconductor is n-type with

$$N_D \simeq n = n_i e^{(E_F - E_i)/kT} = n_i e^{E_G/4kT} = (1.18 \times 10^{10}) e^{1.12/0.1036}$$

$$= 5.85 \times 10^{14}/\text{cm}^3$$

From Fig. 3.5 one deduces $\mu_n \simeq 1350$ cm^2/V-sec, and therefore

$$\rho = \frac{1}{q\mu_n N_D} = \frac{1}{(1.6 \times 10^{-19})(1.35 \times 10^3)(5.85 \times 10^{14})} = 7.91 \ \Omega\text{-cm}$$

(Making use of the Fig. 3.7 resistivity plot one concludes $\rho \simeq 8 \ \Omega$-cm.)

(b) The electron must possess a kinetic energy at least equal to the potential energy hill between the p and n sides of the material, or

$$\text{K. E.}_{\text{min}} = E_c(-W/2) - E_c(W/2) = E_i(-W/2) - E_i(W/2) = E_G/2$$

(c) The V versus x relationship must have the same functional form as the "upsidedown" of E_c (or E_i or E_v) versus x. Thus, with V set equal to zero at $x = -\infty$,

(d) The electric field $\mathscr{E}$ is proportional to the slope of the bands. Thus

(e) The semiconductor is concluded to be in equilibrium because the Fermi level is invariant as a function of position.

(f) $\boxed{J_N = 0}$ and $\boxed{J_P = 0}$. The net electron and hole currents are always identically zero everywhere under equilibrium conditions.

(g) $\boxed{\text{Yes}}$, there is a drift current at $x = 0$. Since $J_{N|\text{drift}} = q\mu_n n\mathscr{E}$, and since both n and $\mathscr{E}$ are nonzero at $x = 0$, $J_{N|\text{drift}} \neq 0$ at $x = 0$. The drift component of the current has the same direction as the electric field—$J_{N|\text{drift}}$ is in the $\boxed{-x \text{ direction}}$.

(h) $\boxed{\text{Yes}}$, there is a diffusion current at $x = 0$. Since $n = n_i \exp[(E_F - E_i)/kT]$, n clearly increases in going from $x = -W/2$ to $x = W/2$. Thus $dn/dx \neq 0$ and $J_{N|\text{diff}} = qD_N(dn/dx) \neq 0$ at $x = 0$. With $dn/dx > 0$ at $x = 0$, $J_{N|\text{diff}}$ flows in the $\boxed{+x \text{ direction}}$. (The diffusion component of the current must of course cancel the drift component of the current so that $J_N = J_{N|\text{drift}} + J_{N|\text{diff}} = 0$ under equilibrium conditions.)

EXERCISE 3.2

Q: In Sample Problem No. 2 of Subsection 3.5.2, nonpenetrating illumination of a semiconductor bar was found to cause a steady state, excess-hole concentration of $\Delta p_n(x) = \Delta p_{n0} \exp(-x/L_P)$.

Given the prevailing low level injection conditions, and noting that $p = p_0 + \Delta p$, we can therefore state

$$n \simeq n_0$$

$$p = p_0 + \Delta p_{n0} e^{-x/L_P}$$

for the illuminated sample.

(a) Making use of Eqs. (3.61), establish relationships for F_N and F_P in the illuminated bar.

(b) Show that F_P is a linear function of x at points where $\Delta p_n(x) \gg p_0$.

(c) Using the results of parts (a) and (b), sketch the energy band diagrams describing the semiconductor bar of Sample Problem No. 2 under equilibrium and illuminated steady state conditions. (Assume $\mathscr{E} = 0$ in the illuminated bar.)

(d) Is there a hole current in the illuminated bar under steady state conditions? Explain.

(e) Is there an electron current in the illuminated bar under steady state conditions? Explain.

A: (a) Since $n \simeq n_0$, it follows that $F_N \simeq E_F$. Likewise, substituting the above p-expression into Eq. (3.61b), we conclude

$$F_P = E_i - kT \ln(p/n_i) = E_i - kT \ln[p_0/n_i + (\Delta p_{n0}/n_i)e^{-x/L_P}]$$

(b) If $\Delta p_n(x) \gg p_0$, then $(\Delta p_{n0}/n_i)\exp(-x/L_P) \gg p_0/n_i$ and

$$F_P \simeq E_i - kT \ln[(\Delta p_{n0}/n_i)e^{-x/L_P}]$$

or

$$F_P = E_i - kT \ln(\Delta p_{n0}/n_i) + kT(x/L_P)$$

(c) We know from Sample Problem No. 2 that $\Delta p_{n0} = 10^{10}/\text{cm}^3$, $n_i = 10^{10}/\text{cm}^3$, and $p_0 = n_i^2/N_D = 10^5/\text{cm}^3$. Thus

(i) Near $x = 0$, $\Delta p_n(x) \gg p_0$ and F_P is a linear function of x.

(ii) At $x = 0$, $\Delta p_{n0} = n_i$ and we deduce from the part (b) result that $F_P = E_i$.

(iii) For large x, F_P eventually approaches $F_N = E_F$.

(iv) $F_N - E_i \simeq E_F - E_i = kT \ln(N_D/n_i) = 0.30$ eV.

Utilizing the above information one concludes

Equilibrium

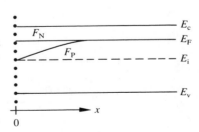

Steady State Illuminated

(d) Assuming $p \neq 0$, it follows from Eq. (3.65a) that there will be a hole current whenever $dF_P/dx \neq 0$. There is obviously a hole current in the illuminated bar near $x = 0$.

(e) Appearances can sometimes be deceiving; $J_N \neq 0$ near $x = 0$! One might conclude from the part (c) result that $dF_N/dx = 0$ and therefore $J_N = 0$. However, under steady state conditions one must have J (total current) $= J_N + J_P = $ constant at all points in the bar. Since no current leaves the bar at $x = 0$, J moreover must vanish at $x = 0$. Thus $J_N(x) = -J_P(x)$ and we know $J_P \neq 0$ near $x = 0$. The apparent discrepancy here stems from the fact that J_N is proportional to both n and ∇F_N. Because the majority carrier electron concentration is much larger than the minority carrier hole concentration, dF_N/dx must be correspondingly smaller than dF_P/dx. The slope in F_N simply cannot be detected by inspecting the energy band diagram.

Appendix B
Volume Review Problem Sets and Answers

The following problem sets were designed assuming a knowledge — at times an integrated knowledge — of the subject matter in the three chapters of Volume I. The sets could serve as a review or as a means of evaluating the reader's mastery of the subject. If the reader is well prepared, each problem set should take approximately 40 minutes to complete under "closed-book" conditions. An answer key is included at the end of the problem sets.

NOTE: (1)   In answering multiple choice questions, choose the *best available* answer: select only *one* answer per question.

              (2)   For numerical computations employ the approximate parametric values listed in Table 3.2.

PROBLEM SET A

I. General Information

(1) As shown below, a cubic unit cell has one atom centered halfway down each of the vertical edges of the cell *and* one atom positioned in the middle of the top and bottom faces of the cell. How many atoms are there within the cell?

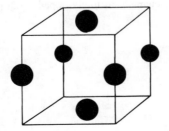

(a) 4

(b) 2

(c) 1

(d) 1/2

(2) What is the name of the *lattice* characterized by the unit cell in question #1?

(a) Simple cubic

(b) bcc

(c) fcc

(d) Diamond

(3) Determine the Miller indices of the plane pictured below.

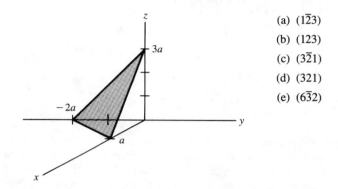

(a) $(1\bar{2}3)$

(b) (123)

(c) $(3\bar{2}1)$

(d) (321)

(e) $(6\bar{3}2)$

(4) Which of the following directions is NOT perpendicular to the [100] direction?

(a) $[0\bar{1}1]$

(b) $[032]$

(c) $[001]$

(d) $[\bar{1}00]$

(5) In a particular semiconductor the probability of electrons occupying states at an energy kT above the bottom of the conduction band is e^{-10}. Determine the position of the Fermi level in the given material.

(a) $E_F = E_c$

(b) $E_c - E_F \simeq 9kT$

(c) $E_c - E_F = 10kT$

(d) $E_F = E_c + kT$

(6) The dominant carrier scattering mechanism in lowly doped (high-purity) silicon at room temperature is

(a) Carrier-carrier scattering

(b) Lattice scattering

(c) Impurity ion scattering

(d) Piezoelectric scattering

(7) Which of the following sketches best describes the D_N versus N_D dependence of electrons in silicon at room temperature?

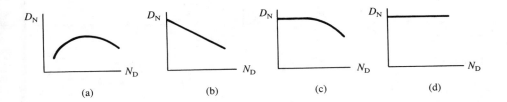

(a)   (b)   (c)   (d)

(8) Using the *bonding* model, indicate how one visualizes a donor.

(9) Using the energy band diagram, indicate how one would visualize the freeze-out of electrons on donor sites as $T \rightarrow 0$ K.

(10) The resistivity of *n*-type material is typically smaller than the resistivity of comparably doped *p*-type material. Explain why.

(11) Using the energy band diagram, indicate how one visualizes *recombination* via R–G centers.

(12) Define "drift velocity."

II. Interpretation of Energy Band Diagrams

A silicon device maintained at 300 K is characterized by the energy band diagram shown below. Use the cited energy band diagram in answering Questions 13 to 20.

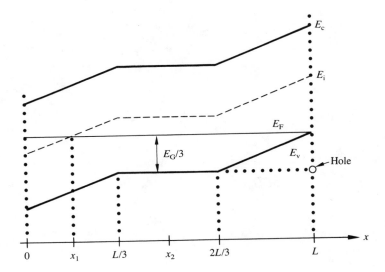

(13) The electrostatic potential (V) inside the semiconductor is as sketched below.

     (a)   (b)   (c)   (d)   (e)

(14) The electric field $(\mathscr{E})$ is as sketched below.

     (a)   (b)   (c)   (d)   (e)

(15) Do equilibrium conditions prevail?

(a) Yes

(b) No

(c) Can't be determined

(16) The semiconductor is degenerate

(a) Near $x = 0$

(b) For $L/3 \le x \le 2L/3$

(c) Near $x = L$

(d) Nowhere

(17) At $x = x_2$, $p = $?

(a) $7.63 \times 10^6/\text{cm}^3$

(b) $1.31 \times 10^{13}/\text{cm}^3$

(c) $10^{10}/\text{cm}^3$

(d) $1.72 \times 10^{16}/\text{cm}^3$

(18) The electron current density (J_N) flowing at $x = x_1$ is

(a) Zero

(b) $\mu_n n_i E_G/L$

(c) $-\mu_n n_i E_G/L$

(d) $D_N[n(x_2) - n(0)]/L$

(19) The hole drift current density $(J_{P|drift})$ flowing at $x = x_1$ is:

(a) Zero

(b) $\mu_p n_i E_G / L$

(c) $-\mu_p n_i E_G / L$

(d) $q \mu_p N_D (kT/q) / L$

(20) The kinetic energy of the hole shown on the diagram is

(a) E_v

(b) $-E_G / 3$

(c) $E_G / 3$

(d) Zero

III. Problem Solving

The semi-infinite n-type silicon bar pictured below is subject to a perturbation such that $\Delta p_n(0) = \Delta p_{n0} > 0$. The bar is maintained at room temperature, steady state conditions prevail, and $N_D = 10^{14}/cm^3$ for all x. The bar is special in that all R–G centers have been removed from the $0 \le x \le L$ region of the bar; i.e., R–G center recombination–generation (thermal R–G) is negligible for $0 \le x \le L$. There are no "other processes," including photogeneration, occurring anywhere inside the bar.

(21) Determine n_0, the equilibrium electron concentration.

(a) $10^{10}/cm^3$

(b) $10^{14}/cm^3$

(c) $10^6/cm^3$

(d) $n_0 \simeq 10^6/cm^3 + \Delta p_{n0}$

(22) If $\Delta p_{n0} = 10^{14}/cm^3$, would low level injection conditions prevail?

(a) Yes

(b) No

(c) Can't determine

(23) to (25) Assuming Δp_{n0} is such that low level injection conditions prevail, show that

$$\Delta p_n(L) = \Delta p_{n0} \frac{L_P}{L + L_P}$$

L_P is the minority carrier diffusion length in the $x \geq L$ portion of the bar.
NOTE: Both $\Delta p_n(x)$ and $d\Delta p_n(x)/dx$ must be continuous at $x = L$.

PROBLEM SET B

I. General Information

(1) As shown below, a cubic unit cell has one atom centered halfway down each of the vertical edges of the cell. How many atoms are there within the unit cell?

(a) 4

(b) 2

(c) 1

(d) ½

(2) What is the name of the lattice characterized by the unit cell in question #1?

(a) Simple cubic

(b) bcc

(c) fcc

(d) Diamond

(3) In a simple cubic lattice with lattice constant a, the density of atoms (number per unit area) whose centers lie on a (100) surface is

(a) $1/a$

(b) $1/a^2$

(c) $4/a^2$

(d) a^2

(4) Determine the Miller indices of the plane pictured below.

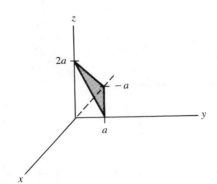

(a) $(\bar{1}12)$

(b) (112)

(c) $(\bar{2}21)$

(d) (221)

(5) The freeze-out of electrons on donor sites as $T \rightarrow 0$ K is best characterized by

(a)   (b)   (c)   (d)

(6) The illustration below is likely to depict the situation inside an n-type semiconductor at approximately what temperature?

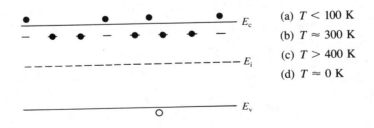

(a) $T < 100$ K

(b) $T \approx 300$ K

(c) $T > 400$ K

(d) $T \approx 0$ K

(7) Consider the sketch of electron density versus temperature for an n-type silicon sample shown below.

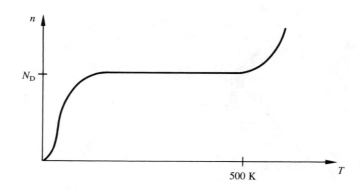

At $T > 500$ K, the sample is said to be maintained in what temperature region?

(a) Extrinsic

(b) Intrinsic

(c) Degenerate

(d) Freeze-out

(8) What is the probability that a state kT below the Fermi level will be occupied by a hole?

(a) Zero

(b) 0.269

(c) 0.368

(d) 0.632

(e) 0.731

(9) A silicon sample is doped with 10^{16} donors *and* 5×10^{15} acceptors per cm^3. At room temperature, the hole density is

(a) $5 \times 10^{15}/\text{cm}^3$

(b) $1 \times 10^{16}/\text{cm}^3$

(c) $2 \times 10^{16}/\text{cm}^3$

(d) $2 \times 10^4/\text{cm}^3$

(e) $1 \times 10^4/\text{cm}^3$

(f) $5 \times 10^3/\text{cm}^3$

(10) Which sketch best describes the μ_n vs. N_D dependence of electrons in silicon?

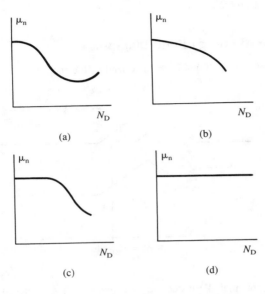

(a)

(b)

(c)

(d)

(11) It is determined that $\mu_p = 500$ cm^2/V-sec at 300 K. If the semiconductor is nondegenerate, what is D_P?

(a) 8.13×10^{19} cm^2/sec

(b) 13 cm^2/sec

(c) 26 cm^2/sec

(d) 1.92×10^4 cm^2/sec

(12) A p-type silicon sample is uniformly doped with $N_A = 10^{15}$/cm^3 and uniformly illuminated such that $\Delta n = \Delta p = 10^{14}$/cm^3. Calculate the resistivity of the illuminated sample. Assume $\mu_n = 1350$ cm^2/V-sec and $\mu_p = 460$ cm^2/V-sec.

(a) 9.75 ohm-cm

(b) 10.5 ohm-cm

(c) 13.6 ohm-cm

(d) 46.3 ohm-cm

(13) A Si sample is uniformly doped with $N_D = 10^{15}$/cm^3 donors. Within a small region of the sample $n = 10^{14}$/cm^3 and $p = 10^4$/cm^3. In this region

(a) Generation rate = recombination rate = zero

(b) Generation rate > recombination rate

(c) Generation rate < recombination rate

(d) Can't tell from the problem

II. Interpretation of Energy Band Diagrams

A semiconductor maintained at 300 K is characterized by the energy band diagram shown below.

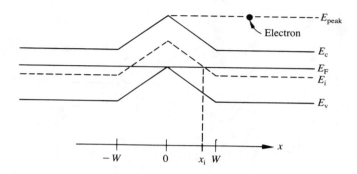

(14) The electrostatic potential (V) inside the semiconductor is as sketched below.

(15) The electric field ($\mathscr{E}$) is as sketched below.

(16) Do equilibrium conditions prevail?

(a) Yes

(b) No

(c) Can't be determined

(17) The semiconductor is degenerate at

(a) $-W \leq x \leq W$

(b) Near $x = 0$

(c) For all x

(d) Nowhere

(18) On a semilog plot the electron concentration is roughly as sketched below.

     (a)   (b)   (c)   (d)

(19) The electron drift current density $(J_{N|drift})$ flowing at $x = x_i$ is

(a) Zero

(b) $\mu_n n_i [E_{peak} - E_c(W)]/W$

(c) $-\mu_n n_i [E_{peak} - E_c(W)]/W$

(d) $\mu_n N_D [E_{peak} - E_c(W)]/W$

(e) $-\mu_n N_D [E_{peak} - E_c(W)]/W$

(20) The hole diffusion current density $(J_{P|diff})$ flowing at $x = x_i$ is

(a) Zero

(b) $\mu_p n_i [E_{peak} - E_c(W)]/W$

(c) $-\mu_p n_i [E_{peak} - E_c(W)]/W$

(d) $qD_P N_D/W$

(e) $-qD_P N_D/W$

(21) The kinetic energy of the electron shown on the diagram is

(a) $E_{peak} - E_c(W)$

(b) $E_c(W) - E_{peak}$

(c) Zero

(d) $E_G + E_{peak} - E_c(W)$

III. Problem Solving

A uniformly doped n-type silicon bar of length L is maintained at room temperature under steady state conditions such that $\Delta p_n(0) = \Delta p_{n0} > 0$ and $\Delta p_n(L) = 0$. $N_D = 10^{15}$ cm^{-3}, $\Delta p_{n0} \ll n_0$, and there are no "other processes" (including photogeneration) occurring inside the bar.

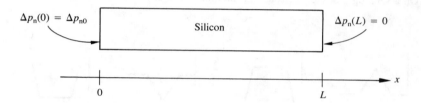

(22) Determine p_0, the equilibrium hole concentration.

(a) $p_0 \approx 10^{10}/\text{cm}^3$

(b) $p_0 \approx 10^{15}/\text{cm}^3$

(c) $p_0 \approx 10^5/\text{cm}^3$

(d) Not given

(23) Does low level injection exist under steady state conditions?

(a) Yes

(b) No

(c) Can't tell

(24) The equation (simplest form possible) one must solve to determine $\Delta p_n(x)$ is

(a) $0 = D_P \dfrac{d^2 \Delta p_n}{dx^2}$

(b) $\dfrac{d \Delta p_n}{dt} = \dfrac{\Delta p_n}{\tau_p}$

(c) $\dfrac{\partial \Delta p_n}{\partial t} = D_P \dfrac{\partial^2 \Delta p_n}{\partial x^2} - \dfrac{\Delta p_n}{\tau_p}$

(d) $\dfrac{\partial p_{\mathrm{n}}}{\partial t} = D_{\mathrm{P}}\dfrac{\partial^2 \Delta p_{\mathrm{n}}}{\partial x^2}$

(e) $0 = D_{\mathrm{P}}\dfrac{d^2 \Delta p_{\mathrm{n}}}{dx^2} - \dfrac{\Delta p_{\mathrm{n}}}{\tau_{\mathrm{p}}}$

(25) The general solution for $\Delta p_{\mathrm{n}}(x)$ will be $(L_{\mathrm{P}}^2 \equiv D_{\mathrm{P}}\tau_{\mathrm{p}})$

(a) $A \exp(-x/L_{\mathrm{P}}) + B \exp(x/L_{\mathrm{P}})$

(b) $A \exp(-t/\tau_{\mathrm{p}})$

(c) $A + Bx$

(d) $A \exp(-x/L_{\mathrm{P}})$

ANSWERS—SET A

(1) b... $1/2 \times$ (2 face atoms) $+ 1/4 \times$ (4 edge atoms) $= 2$

(2) b

(3) e... 1, -2, 3   — intercepts
          1, $-1/2$, 1/3   — 1/intercepts
          $(6\bar{3}2)$ — indices

(4) d... Any direction with a nonzero first integer will *not* be perpendicular to the [100] direction.

(5) b... From the statement of the problem we conclude

$$f(E_{\mathrm{c}} + kT) = \frac{1}{1 + e^{(E_{\mathrm{c}}+kT-E_{\mathrm{F}})/kT}} = e^{-10}$$

$$E_{\mathrm{c}} - E_{\mathrm{F}} = kT[\ln(e^{10} - 1) - 1] \approx 9kT$$

(6) b

(7) c... For nondegenerate dopings $D_{\mathrm{N}} = (kT/q)\mu_{\mathrm{n}}$ and μ_{n} versus N_{D} has the form of sketch (c). (See Fig. 3.5.)

(8)

DONOR
with 5
valence
electrons

(9)

E_c
E_D

E_v

(10)

$$\rho = \frac{1}{q\mu_n N_D} \quad \ldots n\text{-type}$$

$$\rho = \frac{1}{q\mu_p N_A} \quad \ldots p\text{-type}$$

Typically, $\mu_p < \mu_n$ for a given semiconductor and temperature. Thus $\rho(p\text{-type})$ is greater than $\rho(n\text{-type})$ if $N_D = N_A$.

(11)

E_c

E_T

E_v

(12) "Drift velocity" is the average electron or hole velocity resulting from an applied electric field. (One might also mention that at low fields $v_d = \mu\mathscr{E}$, where μ is the carrier mobility.)

(13) d . . . V has the same shape as the "upsidedown" of the bands.

(14) e . . . $\mathscr{E}$ is proportional to the slope of the bands.

(15) a . . . E_F is invariant with position.

(16) c . . . E_F approaches E_v near $x = L$.

(17) b . . .

$$E_i - E_F = E_G/2 - E_G/3 = E_G/6$$
$$p = n_i e^{(E_i - E_F)/kT} = 10^{10} e^{1.12/(6 \times 0.026)} = 1.31 \times 10^{13}/\text{cm}^3$$

(18) a . . . Under equilibrium conditions $J_N = 0$.

(19) b . . .

$$J_{P|drift} = q\mu_p p \mathscr{E}$$

At $x = x_1$

$$p = n_i \qquad (E_i = E_F)$$

$$\mathscr{E} = \frac{1}{q}\frac{dE_i}{dx} = \frac{1}{q}\frac{\Delta E_i}{\Delta x} = \frac{1}{q}\frac{(E_G/3)}{(L/3)} = \frac{E_G}{qL}$$

$$J_{P|drift} = \mu_p n_i E_G/L$$

(20) c . . . K. E. $= E_v(L) - E_{hole} = E_G/3$

(21) b . . . $n_0 \simeq N_D = 10^{14}/cm^3$

(22) b . . . For low level injection conditions to prevail one requires $\Delta p_{n0} \ll n_0$. Here $\Delta p_{n0} = n_0$.

(23) to (25)

For $0 \le x \le L$ . . . $0 = D_P \dfrac{d^2\Delta p_n}{dx^2}$ and $\Delta p_n(x) = A + Bx$

but $\Delta p_n(0) = \Delta p_{n0}$, and $\Delta p_n(x) = \Delta p_{n0} + Bx$

For $x \ge L$ . . . $0 = D_P \dfrac{d^2\Delta p_n}{dx^2} - \dfrac{\Delta p_n}{\tau_p}$

$$\Delta p_n(x') = A'e^{-x'/L_P} + B'e^{x'/L_P} \qquad \ldots \text{general solution} \\ [x' = 0 \text{ at } x = L]$$

but $\Delta p_n(\infty) = 0$, requiring $B' = 0$

$$\Delta p_n(x') = A'e^{-x'/L_P}$$

Next applying the continuity requirements at $x = L$ (or $x' = 0$) yields

$$\Delta p_{n0} + BL = A' \qquad (\Delta p_n \text{ continuous})$$
$$B = -A'/L_P \qquad (d\Delta p_n/dx \text{ continuous})$$

Thus

$$B = -\Delta p_{n0}/(L + L_P)$$

and

$$\Delta p_n(L) = \Delta p_{n0} - \Delta p_{n0}\frac{L}{L + L_P} = \Delta p_{n0}\frac{L_P}{L + L_P}$$

ANSWERS—SET B

(1) c . . . $1/4 \times$ (4 edge atoms) = 1 atom

(2) a

(3) b . . .

—1/4 atom each corner
density = $(1/4 \times 4)/a^2 = 1/a^2$

(4) c . . . $-1, 1, 2$ — intercepts
$-1, 1, 1/2$ — 1/intercepts
$(\bar{2}21)$ — indices

(5) b . . . Only sketches (b) and (c) show donor sites; as $T \rightarrow 0$ K the sites become filled with electrons.

(6) a . . . The figure shows the electrons on some of the donor sites. This occurs as the temperature is reduced toward $T = 0$ K. Thus $T < 100$ K is the best available answer. (At $T = 0$ K *all* of the donor sites would be filled.)

(7) b . . . See Fig. 2.22.

(8) b . . . The probability that a state is occupied by a *hole* is given by $1 - f(E)$. Here $E = E_F - kT$.

$$1 - f(E_F - kT) = \frac{1}{1 + e^{-(E_F - kT - E_F)/kT}} = \frac{1}{1 + e^1} = 0.269$$

(9) d . . . $n \simeq N_D - N_A = 5 \times 10^{15}/cm^3$

$$p = n_i^2/n = (10^{20})/(5 \times 10^{15}) = 2 \times 10^4/cm^3$$

(10) c . . . See Fig. 3.5.

(11) b . . . $D_P = (kT/q)\mu_p = (0.026)(500) = 13\ cm^2/sec$

(12) a . . .

$$\rho = \frac{1}{q(\mu_n n + \mu_p p)}$$

$$p_0 \simeq N_A = 10^{15}/\text{cm}^3; \qquad n_0 = n_i^2/p_0 = 10^5/\text{cm}^3$$

$$p = p_0 + \Delta p = 1.1 \times 10^{15}/\text{cm}^3$$

$$n = n_0 + \Delta n \simeq 10^{14}/\text{cm}^3$$

$$\rho \simeq \frac{1}{(1.6 \times 10^{-19})\,[(1350)\,(10^{14}) + (460)\,(1.1 \times 10^{15})]}$$

$$= 9.75 \text{ ohm-cm}$$

(13) b ... Since $n_0 \simeq N_D = 10^{15}/\text{cm}^3$ and $p_0 = n_i^2/n_0 \simeq 10^5/\text{cm}^3$, there is a carrier deficit inside the cited region of the sample. Whenever there is a carrier deficit the generation rate is greater than the recombination rate.

(14) b ... V has the same shape as the "upsidedown" of the bands.

(15) c ... $\mathcal{E}$ is proportional to the slope of the bands.

(16) a ... E_F is invariant with position.

(17) b ... E_F approaches E_v near $x = 0$.

(18) c ...

$$n = n_i e^{(E_F - E_i)/kT} \qquad \text{(in nondegenerate regions)}$$

From the diagram we note that —

(i) $E_F - E_i = $ constant, $n = $ constant for $x < -W$ and $x > W$.

(ii) $E_F - E_i$ becomes negative and n is at a minimum near $x = 0$.

Only sketch (c) has the proper general shape.

(19) c ...

$$J_{N|drift} = q\mu_n n \mathcal{E}$$

$$\text{At } x = x_i$$

$$n = n_i \qquad (E_F = E_i)$$

$$\mathcal{E} = \frac{1}{q}\frac{dE_i}{dx} = \frac{1}{q}\frac{\Delta E_i}{\Delta x} = -\frac{1}{q}\frac{[E_{peak} - E_c(W)]}{W}$$

$$J_{N|drift} = -\mu_n n_i [E_{peak} - E_c(W)]/W$$

(20) b ... Under equilibrium conditions $J_{P|drift} + J_{P|diff} = 0$. Therefore,

$$J_{P|diff} = -J_{P|drift} = -q\mu_p p \mathcal{E}$$

$$\text{At} \quad x = x_i,$$

$$p = n_i \quad (E_F - E_i)$$

$$\mathscr{E} = -\frac{1}{q} \frac{[E_{peak} - E_c(W)]}{W} \quad \dots \text{as noted in (19)}$$

$$J_{P|diff} = \mu_p n_i [E_{peak} - E_c(W)]/W$$

(21) a . . . K. E. $= E_{electron} - E_c(W) = E_{peak} - E_c(W)$

(22) c . . . $n_0 \simeq N_D = 10^{15}/\text{cm}^3$

$$p_0 \simeq n_i^2/n_0 \simeq 10^{20}/10^{15} = 10^5/\text{cm}^3$$

(23) a . . . $\Delta p_{n|max} = \Delta p_{n0} \ll n_0$

(24) e . . . Steady state conditions and the absence of photogeneration allow one to eliminate the $\partial/\partial t$ and G_L terms in the minority carrier diffusion equation [Eq. (3.40b)].

(25) a . . . This is Solution No. 1 cited in Subsection 3.4.3.

Appendix C
List of Symbols

A	area
c_n	electron capture coefficient
c_p	hole capture coefficient
D_N	electron diffusion coefficient (cm^2/sec)
D_P	hole diffusion coefficient (cm^2/sec)
$\mathscr{E}, \mathscr{E}$	electric field
E	energy
E_A	acceptor energy level
E_B	binding energy at dopant (donor, acceptor) sites
E_c	lowest possible conduction band energy
E_D	donor energy level
E_F	Fermi energy or Fermi level
E_G	band gap or forbidden gap energy
E_H	electron binding energy within the hydrogen atom
E_i	intrinsic Fermi level
E_T	trap or R–G center energy level
E_v	highest possible valence band energy
$\mathbf{F}$	force
$\mathscr{F}$	four-point probe correction factor
$f(E)$	Fermi function
F_N	quasi-Fermi level (or energy) for electrons
F_P	quasi-Fermi level (or energy) for holes
$F_{1/2}(\eta)$	Fermi-Dirac integral of order $1/2$
$g_c(E)$	density of conduction band states
G_L	photogeneration rate; number of electron–hole pairs created per cm^3-sec

g_N	net general generation rate of electrons due to "other processes"
g_P	net general generation rate of holes due to "other processes"
$g_v(E)$	density of valence band states
h	Planck's constant
$\hbar$	$h/2\pi$
I	current
I_P	hole current
$I_{P|drift}$	hole current due to drift
$\mathbf{J}, J$	current density (amps/cm^2)
$\mathbf{J}_N, J_N$	electron current density
$\mathbf{J}_{N|diff}$	electron current density due to diffusion
$\mathbf{J}_{N|drift}$	electron current density due to drift
J_{Nx}, J_{Ny}, J_{Nz}	x, y and z direction components of the electron current density
$\mathbf{J}_P, J_P$	hole current density
$\mathbf{J}_{P|diff}$	hole current density due to diffusion
$\mathbf{J}_{P|drift}$	hole current density due to drift
J_{Px}, J_{Py}, J_{Pz}	x, y and z direction components of the hole current density
k	Boltzmann's constant (8.62×10^{-5} eV/K)
K. E.	kinetic energy
K_S	semiconductor (usually Si) dielectric constant
L_N	electron minority carrier diffusion length
L_P	hole minority carrier diffusion length
m_n^*	electron effective mass
m_0	free electron mass
m_p^*	hole effective mass
n	electron carrier concentration (number of electrons/cm^3)
n^+	heavily doped n-type material
N_A	total number of acceptor atoms or sites/cm^3
N_A^-	number of ionized (negatively charged) acceptor sites/cm^3
N_C	effective density of conduction band states
N_D	total number of donor atoms or sites/cm^3
N_D^+	number of ionized (positively charged) donor sites/cm^3
n_i	intrinsic carrier concentration
n_0	equilibrium electron concentration
N_T	number of R$-$G centers/cm^3
N_V	effective density of valence band states
p	hole concentration (number of holes/cm^3)
p^+	heavily doped p-type material

P. E.	potential energy
p_0	equilibrium hole concentration
q	magnitude of the electronic charge (1.6×10^{-19} coul)
r_N	net general thermal recombination rate for electrons
r_P	net general thermal recombination rate for holes
s	probe-to-probe spacing in a four-point probe
t	time
T	temperature
$\boldsymbol{v}$	velocity
V	voltage; electrostatic potential
$\boldsymbol{v}_d, v_d$	drift velocity
Δn	$\Delta n = n - n_0$; deviation in the electron concentration from its equilibrium value
Δn_p	Δn in p-type material
Δp	$\Delta p = p - p_0$; deviation in the hole concentration from its equilibrium value
Δp_n	Δp in n-type material
ε_0	permittivity of free space (8.85×10^{-14} farads/cm)
η	$\eta = (E - E_c)/kT$
η_c	$\eta_c = (E_F - E_c)/kT$
η_v	$\eta_v = (E_v - E_F)/kT$
μ_n	electron mobility
μ_p	hole mobility
ρ	resistivity (ohm-cm) *or* charge density (coul/cm^3)
σ	conductivity
τ_n	electron minority carrier lifetime
τ_p	hole minority carrier lifetime

Index